MANUFACTURING TECHNOLOGY I

Basic Machines and Processes

Manufacturing Technology Volume II

Proposed Contents:

1 Gear Cutting Machines and Processes
2 Honing and Superfinishing Processes
3 Broaching Machines and Operations
4 Advanced Tracer Controlled Copying
5 Electro-Chemical and Laser Machining
6 Unit Construction of Machine Tools
7 Development of Automatic Transfer Lines
8 Mechanical Robots and Automatic Work Control
9 Production Planning
10 Computer Controlled and N.C. Machine Tools
11 Machine Tool Laboratory Work

MANUFACTURING TECHNOLOGY I

Basic Machines and Processes

H C TOWN and H MOORE

B.T.BATSFORD LTD London

First published 1979

ISBN 0 7134 1094 9 (cased edition)
0 7134 1095 7 (limp edition)

Printed and bound by Billing & Sons Ltd,
Guildford, London & Worcester
for the publishers B T Batsford Limited
4 Fitzhardinge Street, London W1H 0AH

CONTENTS

A machine which is symbolic of the current high levels of sophistication in machine tool engineering, a vertical boring and turning mill for the nuclear industry. The machine has a table capacity of 20 ft (6.1 m) and can accept tubular workpieces up to the same height. It has three columns, one mounted on a stationary island in the middle of the chuck and the other two on slides either side of the chuck. The middle column can turn or grind; the outer two can either turn, mill, drill, bore or tap as a fine positioning control is fitted to the chuck. The workpiece weight can be up to 180 t, and the cat walk joining the left-hand column to the central one swings away to permit it to be loaded and unloaded (By courtesy of Industrial Sales Ltd, Wilmslow)

1

THEORY AND PRACTICE OF METAL CUTTING

The earliest cutting material was carbon tool steel, but its use is now restricted to the making of intricate form tools which cannot be ground after hardening. Little surface decarbonization takes place in the furnace; thus the tool will retain its original size after heat treatment. The material is suitable for hand tools such as reamers, taps and dies, as well as for small drills where toughness is important. The hardness of the steel is determined by the carbon content which ranges from 0.7 to 1.4%. Hardening is by quenching between 750 and 800°C in water or oil.

The suitability of a metal-cutting material is decided by its ability to withstand the heat, pressure and abrasion caused by the cutting action of tool and chip. Friction generates heat which causes temperatures as high as 600°C on the cutting edge, and to withstand this heat the material must possess a high 'red-hardness' value. This is defined as the measure of the hardness of a material at elevated temperatures. Chip movement tends to wear the tool, while pressure bearing down on the tool is resisted by the toughness of the material. Modern carbon steels are improved by the addition of some alloying element such as chromium or tungsten.

HIGH-SPEED STEEL

High-speed steel was first introduced in 1900 at the Paris Exhibition and led to a large increase in cutting speeds and more advanced design of machine tools. The steel is classified by its tungsten content, and Table 1 gives the composition of four types. The heat treatment is effected by quenching from about 1,300°C and by then tempering at 500 to 600°C. This tempering is better termed secondary hardness treatment, for, unlike carbon steel, the hardness is increased after treatment. One disadvantage is surface decarburization which takes place in the furnace and leaves scale about 0.005 in deep to be removed by grinding.

Although high-speed steel turning tools have been largely replaced by cemented carbides, other types of high-speed steel cutting tools continue to be manufactured, e.g. drills.

Table 1.1.
Composition percentage of high-speed steels

Tungsten	Carbon	Chromium	Cobalt	Molybdenum	Vanadium
14	0.7	3.4	–	–	–
18	0.75	4	–	–	0.8
18	0.7	4	5	0.5	1
22	–	5	12	0.5	1

STELLITE

Stellite is a non-ferrous alloy and is an alloy of cobalt, chromium and tungsten, which cannot be heat treated but is cast in the required shapes. It possess a high red-hardness value and can only be machined by grinding. Stellite is employed in the form of tips and is specially suitable for blades of inserted milling cutters, as it has a high tensile strength of 80,000 lb/in^2 (5,624 kg/cm^2). Another valuable feature is that it can be welded.

CEMENTED CARBIDES

The term cemented carbides covers a range of cutting tools of which the basic alloy is tungsten carbide. It is a hard compound held in a matrix of cobalt, the carbide being the cutting medium and the cobalt the bond. Because of brittleness the material is always used with tips brazed onto carbon steel shanks or made in the form of 'throw-away' tips. To produce the tips, tungsten and carbide are mixed and heated in the electric furnace to produce hard tungsten carbide. This is reduced to powder form, the cobalt is added and the mixture is pressed into block form. The blocks are then placed in the electric furnace and are sintered in a hydrogen atmosphere. The blocks now resemble graphite and can be cut to shape, so the next process is to place the tips in refractory crucibles and to sinter again but at a much higher temperature. When cool the tips are ready for use for metal cutting.

The original tungsten carbide, while excellent for cutting cast iron, was not so good on steel, because particles of the cuttings tended to build up on the tool. To resist this undesirable feature, tantalum and titanium are added, whilst an alternative to tungsten is to use carbides of molybdenum or titanium.

CARBIDE TOOLS

Carbide tools are available with different grades of carbide to suit the various materials to be cut, and Figure 1.1 shows a range of tools available for different machining operations. The tips must be mounted on the tool

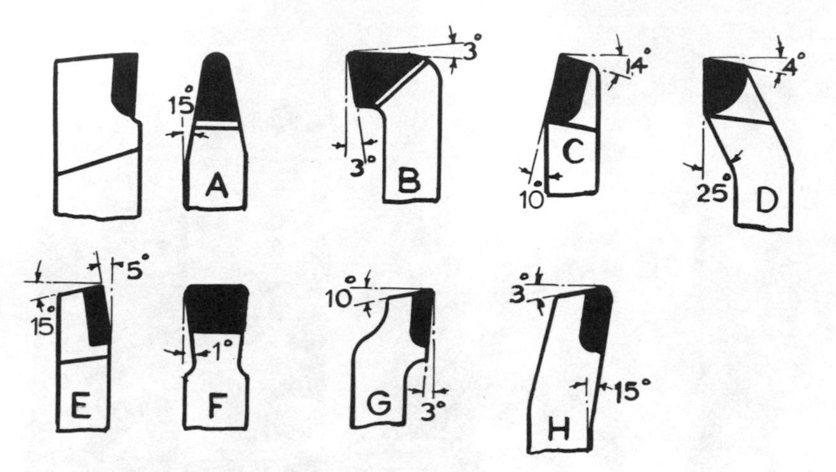

Figure 1.1.
Types of cutting tools with cemented carbide tips:
A round nose turning tool
B turning and facing tool
C bar turning tool
D cranked knife tool
E roughing tool
F finishing
G knife tool
H side roughing tool

shank in such a manner to provide adequate support for the cutting edge, and electronic induction brazing provides the means for attaching the tip to the tool. It is necessary when machining steel to provide means to break up the chip; otherwise it leaves the workpiece in the form of a hot barbed ribbon. Figure 1.2(a) shows a chip curler which causes the cutting to form a spiral and break up, while Figure 1.2(b) shows a tool with a groove ground on the face so that the cutting strikes the back of the groove and breaks up.

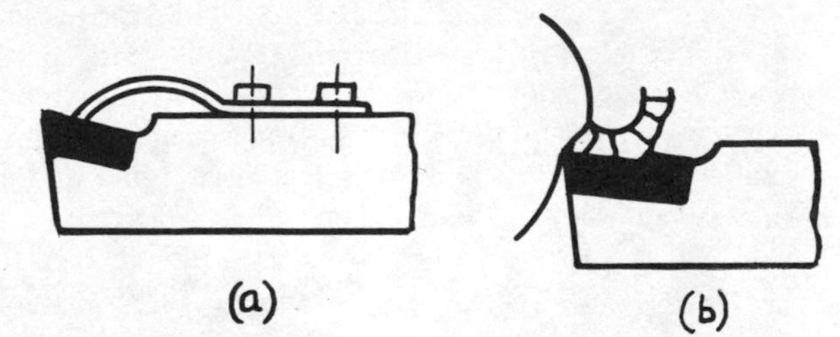

Figure 1.2.
Cutting tools with chip breakers

Because of the cost of grinding carbide tools an alternative to the brazed tip is the 'throw-away' insert as shown on the set of tools used on the Dean, Smith & Grace Ltd numerical control (N-C) lathe shown in Figure 1.3. Triangular and rectangular inserts are indicated, and after one edge is blunted the insert can be turned for a new edge to be presented to the work. In some cases the rectangular insert can be used to give four cutting faces and can then be inverted in the holder, so that eight cutting edges are available, thus reducing the cost. When a new cutting edge is required, it may be necessary to pre-set the tool off the machine, and the gauging fixture of Figure 1.4 is used for this. By means of end buttons the

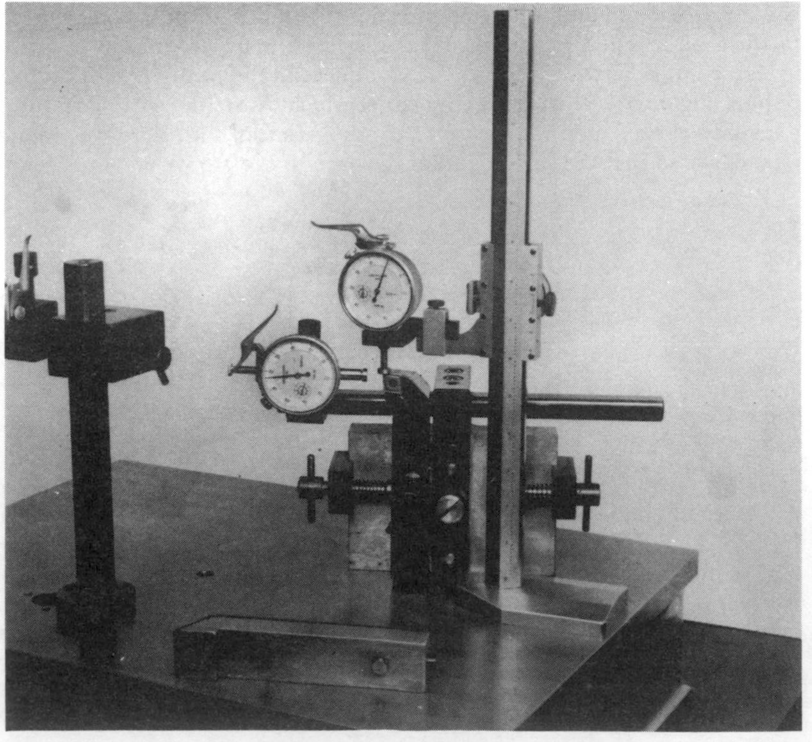

Figure 1.3.
A set of turning tools with 'throw-away' inserts

tool can be set vertically in the fixture, and the position of the tool point in the X and Y axes can be checked by the dial indicators. In general these inserts are set in the holder by a clamp or one retaining screw so that adjustment or replacement is easily carried out by a small socket key.

CERAMIC TOOLS

The main advantage of ceramic tools over carbides lies in their greater hardness and strength at high temperatures; thus higher cutting speeds can be used without loss of tool life. They have, however, little resistance to impact because they are brittle. Ceramics are chemically inert, and, if cratering takes place during machining, it is by mechanized abrasion rather than by chip welding. Because of low heat conductivity, ceramics remain cool at high cutting speeds. Properties of the material vary, although the major constituent is always $A1^{2}0^{3}$, i.e. up to 90% aluminium oxide. Table 1.2 gives a list of suitable cutting speeds for ceramic tools.

Table 1.2.
Suitable cutting speeds for ceramic tools

	Speed (ft/mm) speed (m/min)	
Material	Roughing	Finishing
.1 to .2% carbon steel	600 (180)	1,200 (360)
.2 to .3% carbon steel	500 (150)	900 (273)
.3 to .4% carbon steel	350 (106)	800 (242)
.4 to .6% carbon steel	250 (75)	500 (150)
Cast iron, Brinell 217	750 (226)	1,250 (379)

Commercial brass and aluminium can be machined at almost unlimited speeds depending on the speed range of the machine, the power available and the condition of the machine in which vibration must be at a minimum. In all cases inserts of the 'throw-away' tip type are used, as shown in Figure 1.5, which indicate the shapes available. They are usually set in the holder with an approach angle of 15° and 7° negative top rake. Suitable feeds are from 0.016 to 0.02 in/rev (0.35 to 0.5 mm/rev) with a depth of cut 0.16 to 0.28 in (0.35 to 07 mm).

Figure 1.4.
A gauging fixture for pre-setting turning tools
(By courtesy of Dean, Smith & Grace Ltd, Keighley)

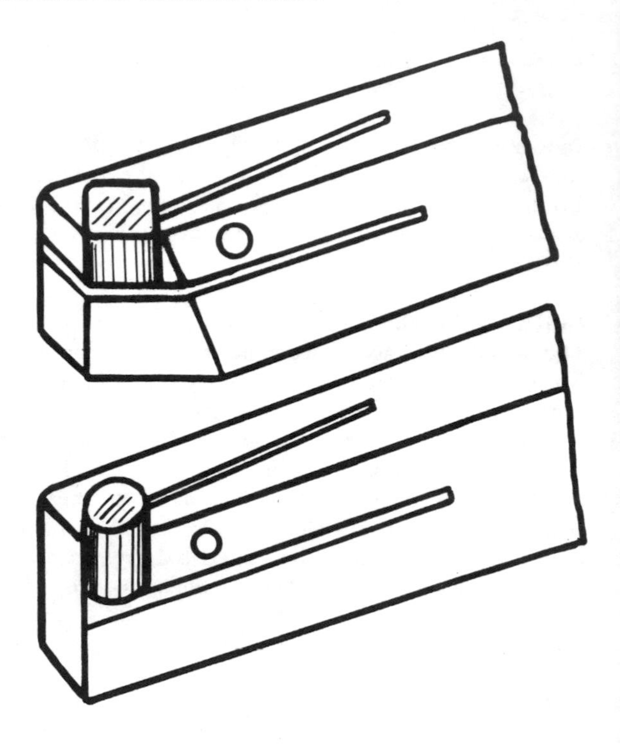

Figure 1.5.
Tool holders with ceramic inserts

COATED CARBIDES

While conventional cemented carbides are well established as the main cutting material in engineering workshops, new developments are still taking place and improved tools in which carbide is still the basic component are available. The improvement comprises a thin layer of hard material bonded on top of the carbide to give either a longer life to the tool or the possibility of using higher cutting speeds.

The Goldmaster process comprises a thin layer 0.5 μm of titanium carbide which is deposited on the base material, followed by a substantial layer of titanium carbo-nitride which bonds with the intermediate layer,

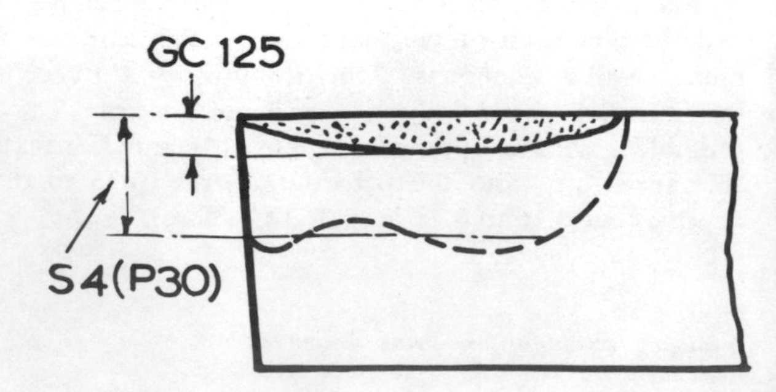

Figure 1.6.
A cutting tool with a GC showing wear reduction

so that the nitride content increases towards the surface. The result is a strata combination which possesses low thermal conductivity, so that the heat generated during cutting is largely contained in the chip and passes away with it. Plastic deformation is drastically reduced, resulting in a much longer life to the cutting edge, so that in certain applications production has been increased by ten times.

GAMMA COATING PROCESS

The word gamma in the term gamma coating (GC) designates the wear-resistant phase in the carbide. Thus in the Sandvik process a coating of fine-grained titanium carbide of very high hardness is used. It is 0.005 mm thick and yet is automatically bonded to the surface of the insert core. Tests have shown it to be extremely wear resistant; Figure 1.6 shows a comparison of the rates of wear of the coated carbide GC with a normal tool S4.

The coated carbide is suitable for the majority of the operations when machining steel and cast iron, and particularly for turning. Figure 1.7 shows a Sandvik tool, but milling operations are also successful at a moderate feed. When turning, the best economy is obtained at a cutting speed 20% higher than when using conventional inserts, and three times the average life of the tool can be expected.

Figure 1.7.
A Sandvik cutting tool with a coated carbide
(By courtesy of Sandvik U.K. Ltd, Halesowen)

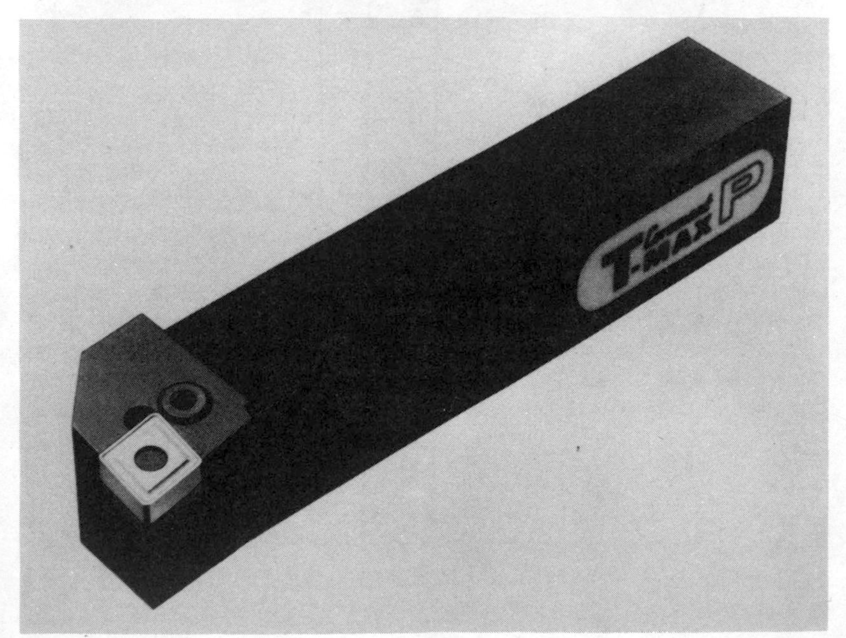

While GC inserts cost somewhat more than ordinary carbide ones, in general because of lower machining costs and a reduction in the number of tools replaced, the tool cost becomes of comparatively subordinate importance through higher productivity. Figure 1.8 is a comprehensive diagram illustrating the relative machining costs and tool life as a function of relative cutting speed.

BZN BONDING MATERIAL

BZN bonding material (General Electric Co. Ltd, London) is produced by bonding a layer 0.5 mm thick of polycrystalline cubic boron nitride to a cemented carbide substrate. Since Borazon CBN material has a hardness value exceeded only by diamond, it provides high resistance to wear and long tool life, while the substrate gives high shock resistance and is

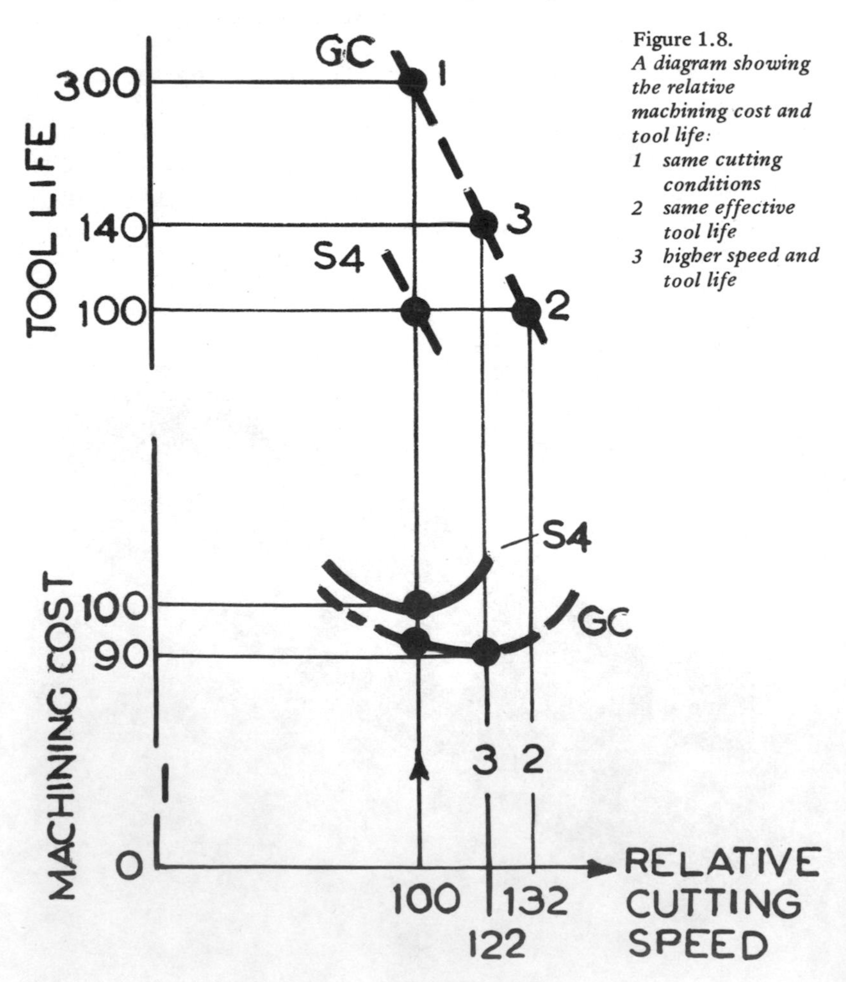

Figure 1.8.
A diagram showing the relative machining cost and tool life:
1 *same cutting conditions*
2 *same effective tool life*
3 *higher speed and tool life*

unaffected at high temperatures. Tool tips are available in round, square and triangular shapes and give best results when machining ferrous metals with a hardness value exceeding 45 Rockwell C.

As an indication of metal removal rates, for turning hardened tool steels, a cut 6.3 mm deep can be taken at a speed of 122 m/min and with a feed rate of 0.25 mm/rev. High-temperature alloys can be machined, and high performances are obtained when turning chilled cast iron and steel rolls with hardness values of up to 47 Rockwell C. With light cuts the surface finish equals that of grinding.

Machining is normally carried out with negative rake cutting, preferably with an angle of 5° to 9° between the tip and workpiece and a lead angle of at least 15°.

DIAMOND TOOLS

The hardest cutting material known is the diamond, but it can only be used for finishing purposes where the cutting forces are small. Suitable operations include turning and boring of components such as aluminium alloy pistons, copper commutators, or milling operations on light alloy crank cases, and diamond honing operations. It is suitable for a wide range of plastic materials and has a limited application in machining cast iron. For the retention of the cutting edge diamonds show an equal superiority over cemented carbides as these do over high-speed steel, the tool life being in the ratio of 3 to 200 in favour of the diamond. As the diamond has no chemical relation to the material to be cut, chip friction is reduced.

Table 1.3.
Cutting speeds for turning and boring

Material	Cutting speed (m/min)	Cutting speed (ft/min)
Aluminium alloys	200 to 300	650 to 1,000
Magnesium alloys	350 to 380	1,000 to 1,250
Bronze	150 to 350	500 to 1,150
Lead-bronze	500 to 600	1,650 to 2,000
White metal	250 to 350	800 to 1,150

Diamond tools provide the best means for turning and boring plastics and will machine Bakelite, Erinoid, Catalin, Tufnol, vulcanite and ebonite with the same high speeds as those listed. Feeds range between 0.008 to 0.004 in/rev (0.2 to 0.1 mm/rev). The cutting depth between 0.004 to 0.025 in (0.1 to 0.6 mm).

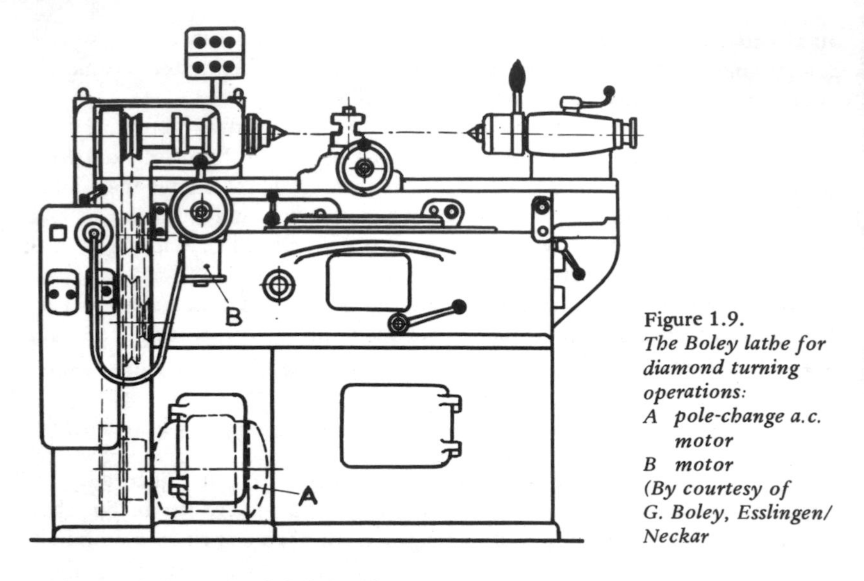

Figure 1.9.
The Boley lathe for diamond turning operations:
A pole-change a.c. motor
B motor
(By courtesy of G. Boley, Esslingen/ Neckar

DIAMOND TURNING LATHES

For efficient machining with diamond tools a special lathe is required with a very-high-speed range, the final drive to the spindle being by endless belt and not through gearing. The motor should be isolated from the machine to prevent vibrations from being transmitted to the tool or work. Low rates of feed should be available, and fine adjustments should be provided for the tool setting.

These features are included in the Boley lathe of Figure 1.9, the headstock drive being by means of a pole-change a.c. motor A, which has three speeds and is mounted on vibration-eliminating pads separate from the machine. An endless belt connects to the spindle and provides three optional speeds of either 750, 1,000 and 1,500 or 1,500, 2,000 and 3,000 rev/min. The spindle is mounted in bronze bearings and is provided with water cooling to ensure a constant temperature.

There are six automatic feeds, the saddle being driven by a screw located between the slideways. By means of motor B, the saddle can be traversed at a rapid speed, and the feed can be engaged at any point. Adjustable stops can be set to terminate the feed at any point, and the tool holder has two locations to allow full use of the tool. A sensitive adjusting device is provided for obtaining a glossy finish with a facet diamond, and the height can be set to scale by a screw. The use of three-jaw chucks should be avoided for they cannot be dynamically balanced and affect the running accuracy of the spindle.

SHAPED DIAMOND TOOLS

Various shapes of cutting edges are used, including diamonds with one cutting edge, diamonds with a circular cutting edge, diamonds with several facets around the tool nose, and specially shaped diamonds for cutting off and profile work. Figure 1.10(a) shows a single-point tool; these usually have no top rake or even a negative rake angle. The clearance angle is kept as small as possible in order not to reduce the tool angle too much. The front adjusting angle is selected between 30 and 45°, while the back adjusting angle is made very small in order to give a polishing action on the finished surface and is actually 1 to 2°. Thus the tool angle is rarely less than 120° except for special operations.

It may be thought that diamond turning is restricted to small components, but this is not the case because drums 24 in in diameter and 48 in long are diamond turned for the computer industry, while comparatively large commutators of copper are machined by single-point tools, as shown in Figure 1.11.

Round nose tools can be considered as cutting edges with an infinite number of facets and are easy to set, but diamonds with from three to seven facets are popular, and this shape is shown in Figure 1.10(b). The advantage is that, after one cutting edge has been blunted, another can be brought into operation. It is advantageous if the tool can be moved

Figure 1.10.
Types of single-point and facet edge diamond tools

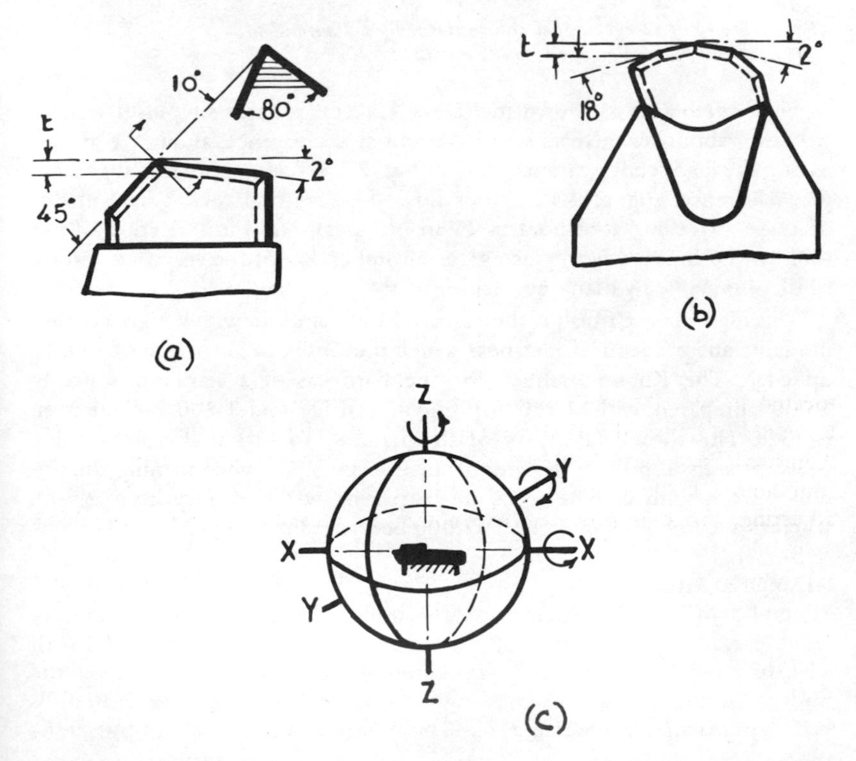

Figure 1.11.
The machining of a large copper commutator by a diamond tool
(By courtesy of L.M. Van Moppes & Sons Ltd)

through three axes as shown in Figure 1.10(c), for this simplifies setting. Rotation about XX affects side rake and side clearance, about YY affects back rake and end clearance and about ZZ affects approach angle and plan clearance. Figure 1.12 shows how the diamond tool is held in one of these swivelling tool holders (Van Moppes), and the reflection of the tool on the finished workpiece gives an indication of the very fine surface finish obtainable by diamond turning.

This high surface finish is the result of the diamond with a high Young's modulus and a chemical inertness which prevents the formation of a built-up edge. The Knoop number for the hardness of a diamond is nearly 7,000 compared with 2,150 for silicon carbide and 1,200 for tungsten carbide. This shows the ability of the diamond to outlast all other cutting tools on a given machining operation. For example, when turning aluminium pistons with carbide tools, resharpening was necessary after 2,500, whereas a diamond completed 107,000 before resharpening.

DIAMOND MILLING

Examples of the process include the milling of the ends of aluminium crank cases using the 30 in (762 mm) cutter head shown in Figure 1.13(a). The diamond holder is shown at B and the method of clamping by the bush A, while Figure 1.13(b) shows a smaller head with two diamonds used for machining bearing caps. The diamond holder is held by screws

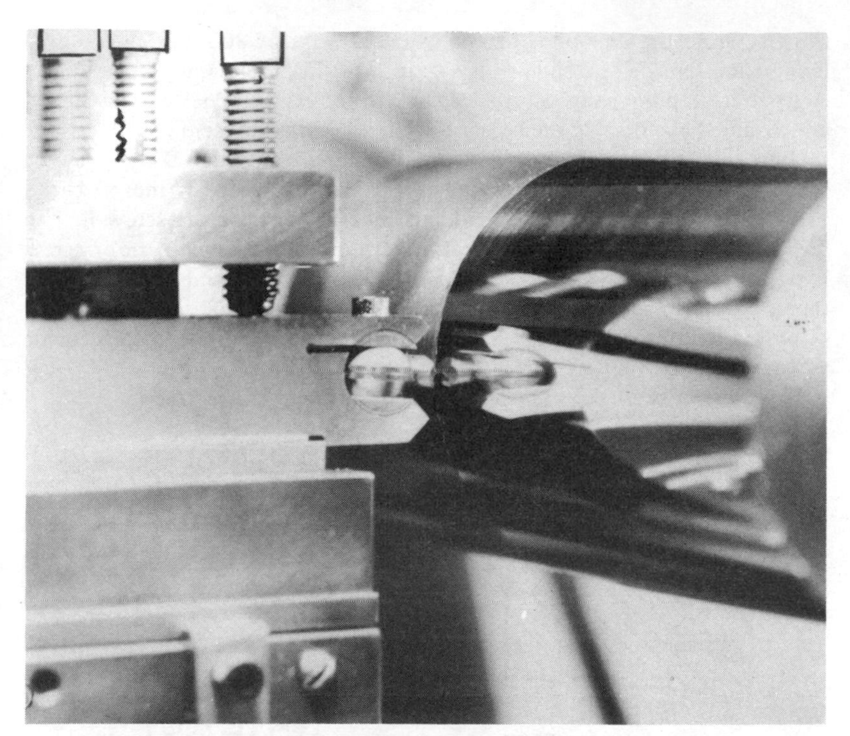

Figure 1.12.
A diamond tool held in swivelling holder

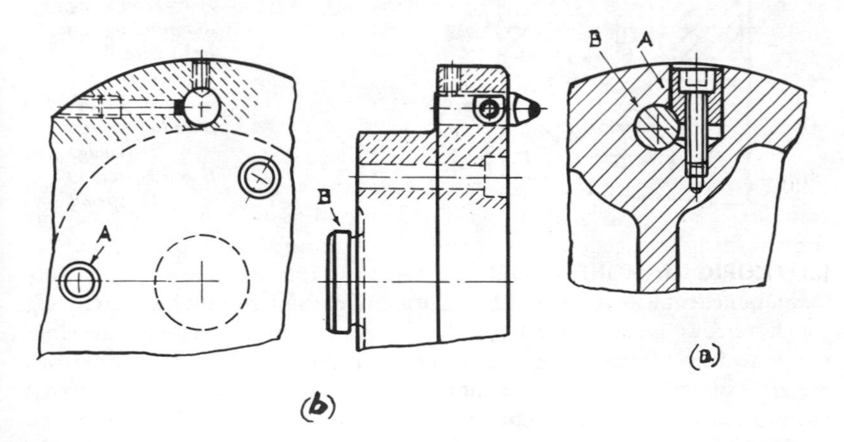

Figure 1.13.
Milling heads with diamond tools for aircraft engine components:
A Bush *B diamond holder*

shown, while the diamonds have a rounded nose of 80°. For fine adjustment a Russian design, shown in Figure 1.14, has the diamond in a holder A so that its point may be set on the correct centre. Coarse adjustments are obtained by rotating the screw B while locking is by screw C.

Fine adjustments are obtained by displacing the wedge D by screw E. When the tool has to be removed, the retaining wedge is moved to its original position, i.e. with its end face bearing against the screw F. The displacement of the tool holder and diamond for a single turn of screw E is given by $\Delta R = SK/2$, where S is the lead of screw and K is the wedge taper, so that $S = 1$ mm and $K = 1$ in 25; then $\Delta R = 0.02$ mm. The spring G puts a loading on the tool holder which ensures uniform displacement when setting the tool.

The subject of diamond boring tools is discussed in Chapter 4.

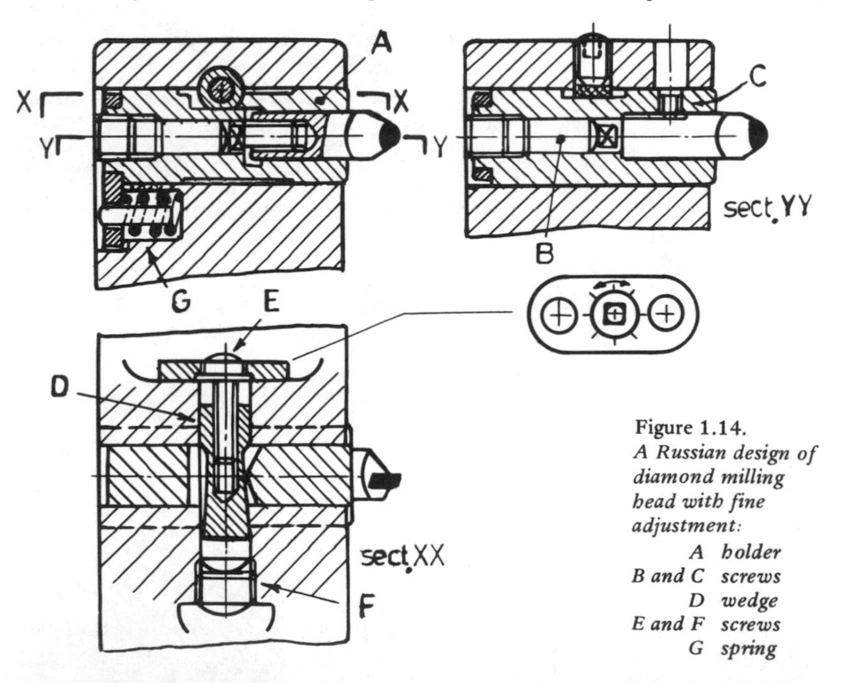

Figure 1.14.
A Russian design of diamond milling head with fine adjustment:
A holder
B and C screws
D wedge
E and F screws
G spring

ISOTROPIC DIAMOND TOOLS

Developments to increase the effective use of carbide tools when machining tough ferrous metals have been described, and in like manner developments to increase the use of diamonds for machining difficult non-ferrous metals have taken place. Prominent in this field the De Beers Industrial Diamond Division Ltd, Ascot, have produced a new material registered as Syndite. This is a synthesized tough intergrown mass of randomly orientated diamond particles in a metal matrix. When bonded onto a tungsten carbide substrate, the result is a powerful composite material

Figure 1.15.
A Syndite tool of isotropic diamond and carbide material

which can be used as an alternative to a conventional diamond tool.

Syndite is produced by sintering together small particles of diamond at temperatures above 1,400°C and at pressures of the order of 60,000 atmospheres. The resulting tool blanks are cut into segments and are shaped for cutting tools, while the tungsten carbide substrate provides a rupture strength of 250 kg/mm^2 where it is most needed, combined with a Young's modulus E of 50,000 kg/mm^2. Between the dense diamond layer and the tungsten carbide there is an intermediate layer of low-modulus metal, because otherwise there is the possibility of mismatched stresses occurring in the diamond-to-carbide bonding regions, (Figure 1.15).

The advantage of this construction is that, while an improperly oriented single-crystal diamond may chip along the cleavage planes, wear of this nature is absent from Syndite diamond tools because the diamond grains are randomly orientated. It is this feature that enables the tools, for either turning or milling, to have a performance which approaches that of a diamond and which in some cases improves upon it. Even though the hardness of Syndite at 8,000 kg/mm^2 is not equal to that of the diamond, the tools can be more rigid, so that a deeper cut can be taken.

Figure 1.16 shows the turning of a copper-coated printing roll 240 mm in diameter and 1,300 mm long. The cutting speed was 754 m/min with a feed of 0.06 mm/rev, and a depth of cut of 0.2 mm. Because this depth was double that possible for a single diamond, the turning was completed in one cut instead of two for the diamond. The tool angles for the Syndite tool were a nose radius of 0.6 mm, a top rake of zero, a clearance angle of 12° and a side cutting angle of 45°. In the other case the turned pistons were of aluminium-silicon alloy at a surface speed of 275 m/min and a depth of cut of 0.5 mm; the number of pistons turned before tool failure was 1,200 for Syndite and 700 for a single diamond.

In regard to milling operations on silicon-aluminium, by using a two-bladed cutter 110 mm in diameter with segments brazed to a steel shank, 25,000 castings were machined. The operation was equivalent to 27,500 m of travel compared with 900 m for a carbide cutter. The rake angle of the tools was zero, the clearance angle was 10° and the depth of cut was variable between 0.7 and 0.9 mm. Table 1.4 gives the relevant information for turning operations on a range of materials.

Table 1.4.
Information required for turning operations

Workpiece material	Speed (m/min)	Feed rate (mm/rev)	Depth of cut (mm)	Nose radius (mm)	Clearance angle (deg)	Rake angle (deg)
Aluminium alloys	600 to 1,200	0.05 to 0.20	0.10 to 0.50	0.20 to 1.00	5 to 10	0 to 5 negative
Copper alloys	500 to 1,250	0.02 to 0.15	0.10 to 1.00	0.25 to 1.00	5 to 25	0 to 15 positive
Bronze	400 to 1,000	0.02 to 0.15	0.10 to 0.50	0.25 to 1.00	5 to 20	0 to 5 negative
Glassfibre-reinforced plastics	150 to 1,000	0.02 to 0.25	0.02 to 0.06	0.80 to 2.50	5 to 20	5 positive to 5 negative
Sintered tungsten carbide	150 to 500	0.02 to 0.05	0.01 to 0.10	1.00 to 3.00	5 to 10	0 to 5 negative
Ceramics	300 to 1,000	0.02 to 0.10	0.01 to 0.10	0.25 to 1.00	5 to 20	5 positive to 5 negative

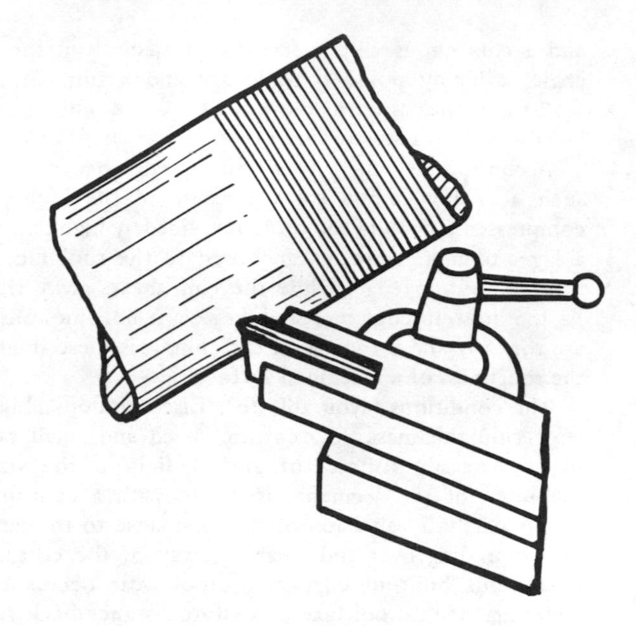

Figure 1.16.
An example of turning copper-coated cylinder with a Syndite tool

CUTTING FORCES AND TOOL ANGLES

The four main groups of factors in metal cutting which affect productivity are surface finish, tool forces and power, tool wear and machine vibrations. Most metal removal is by orthogonal cutting, in which the cutting edge is perpendicular to the direction of cutting and the tool edge is longer than the width of the cut. The alternative method is by an oblique cut not perpendicular to the direction of cutting.

In the machining of metals three types of chips occur; Figure 1.17(a) shows the formation of a discontinuous chip. As the work and tool meet, the point of the latter engages the material, compresses it and causes it to escape upwards along trajectories leading to the free space just above the cutting edge. With additional movement, the flow extends farther ahead of the tool until rupture occurs, and the chip passes off up the face of the tool. With a ductile material successive rupture may not occur,

Figure 1.17.
Types of chips occurring when turning various materials

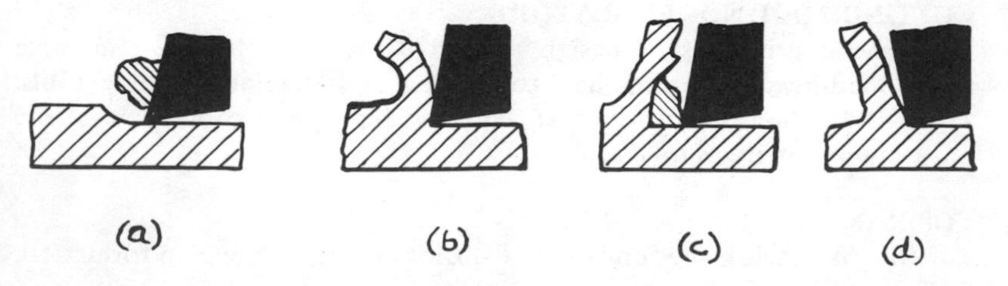

and a continuous chip is formed, particularly if the tool has a large rake angle, a highly polished tool face and a suitable cutting fluid (Figure 1.17(b)). The upward movement of the chip carries the compressed layer with it in a steady flow.

A continuous chip with a built-up edge occurs when machining metals such as a low-carbon steel. Because of the high friction between the compressed layer and the tool face, aided by high temperature and pressure, a layer of metal remains anchored to the tool face, (Figure 1.17(c)) to form a built-up edge, while the chip shears away from it and passes off. As the built-up edge increases in size, it becomes unstable until fragments are torn off and escape with the chip. It is these fragments that constitute the roughness of a machined surface.

The conditions favourable to a discontinuous chip are brittle material, large chip thickness, low cutting speed and small rake angles. This type of chip is easily disposed of, and the finish of the workpiece is good when the pitch of the segments is small. With a continuous chip, tool wear occurs partly by abrasion of the face close to the cutting edge and partly by a rounding over and wearing away of the cutting tool itself. With a continuous built-up edge chip, tool wear occurs both by cupping, or cratering, of the tool face at a short distance back from the cutting edge at the point of contact with the following chip and by abrasion of the tool flank by contact with the fragments of built-up edge which escape with the chips. The crater eventually extends closer and closer to the cutting edge until failure occurs by a splitting off of a portion of the material in this region.

While one particular type of chip is most likely to be encountered with one material, i.e. brittle materials such as cast iron tend to produce discontinuous chips, under different operating conditions of chip thickness, temperature and lubrication of the tool face, there may be a change from one type of chip to another, although the work material remains the same.

The above description refers particularly to lathe, planer and similar single-point tools with straight cutting edges. In cases such as threading and tapping operations, it is essential to reduce friction by a suitable cutting oil and by a high polish on the tool face which prevents the formation of a built-up edge and ensures easy chip flow.

CUTTING TOOL NOMENCLATURE

The general principles of design applied to lathe tools may also, with slight modification, be applied to planing and shaping machine tools. Figure 1.18 shows the various angles on a single-point tool; some of the most important are as follows.

TOP RAKE ANGLE

The top rake angle is the fundamental angle controlling the chip formation.

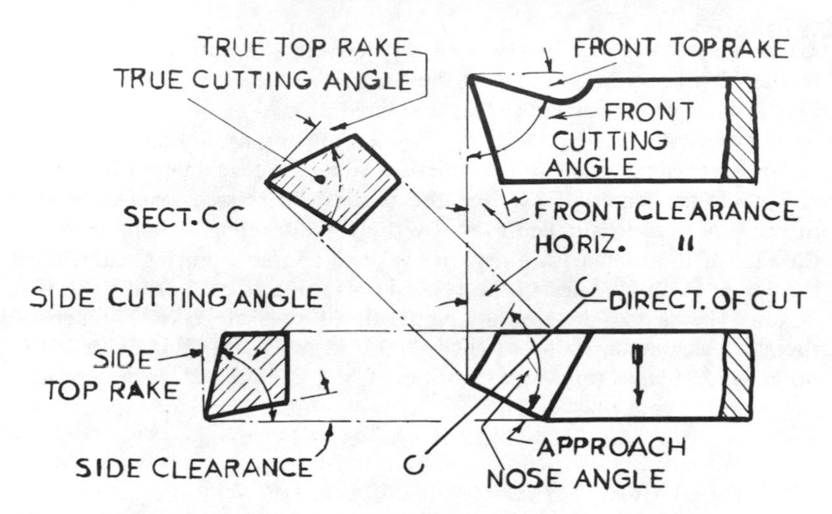

Figure 1.18.
A diagram showing the angles on a single-point cutting tool

It may be defined as the slope of the tool face backwards from the cutting edge. When using a straight cutting edge as in a side turning tool, the chip flows in a right-angled direction to the cutting edge. With round nose tools, the direction of the chip flow depends on the shape of the nose and the relation between the depth of cut and the feed.

CLEARANCE ANGLE
The angle should be just sufficient to prevent the tool flank from rubbing on the work. Excessive clearance merely removes support from the cutting edge. There is no advantage in using clearance angles over 6 to 8°, and in most cases 5° is sufficient.

APPROACH ANGLE
The cutting edge in contact with the work gets longer as the approach angle is increased. Therefore the life of a tool with a large approach angle is longer than one with a small angle. The increased approach angle produces a corresponding increase in the shank thrust; therefore large approach angles should be avoided on work which is long in relation to diameter.

NEGATIVE RAKE MACHINING

The fundamental weakness of cemented carbide and ceramics as a cutting material is the low tensile strength, and to remedy this defect negative rake machining was introduced, (figure 1.17(d)). The negative rake angle varies from 0 to 10°, but the object is to increase the cutting angle, whereas a positive rake weakens the section. Another feature is that the cutting pressure acts further up the tool face, thus holding the brazed tip onto the shank. With a positive rake the cut may tend to pull the tip from its seating. With 'throw-away' tips, the location in the tool holder ensures that a negative rake cutting angle is obtained.

Negative rake single-point tools are used for turning, boring and planing and give excellent results on materials difficult to machine. With positive rake tools the material ahead of the tool is compressed against the face of the tool until it is pulled away by the force reacting at right angles to the face of the tool. With a negative rake this force is much greater, since the forces at right angles to the top rake of the tool are not of a shearing nature. The actual shearing is the result of the side rake. The action, therefore, is a compression of the metal downwards, tending to close the pores, and then a sideways shearing of the compressed chips, giving a very smooth work finish.

MAGNITUDE OF CUTTING FORCES

The shape of a tool depends upon the chip to be removed and on the property of the material that is being cut. The moving chip exerts a force on the tool; the direction that the force is exerted on the tool is downwards, backwards or sideways depending on the tool angles. The total force can be resolved in a single force R, as shown in Figure 1.19. The force can be measured by means of a dynamometer as described in Chapter 2 or can be determined by measuring the power input to the machine-driving motor but subtracting the power taken when the machine is running light.

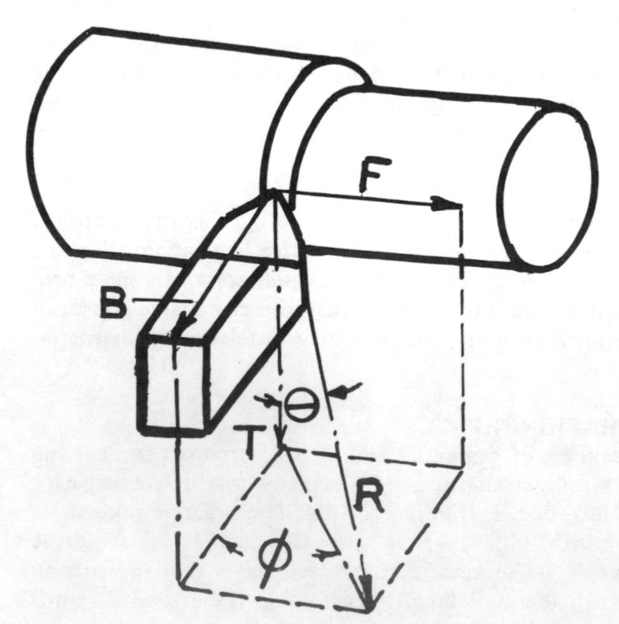

Figure 1.19.
A diagram showing the resultant forces on a cutting tool

A formula used by the Sandvik company is listed and enables the size of the driving motor to be established for any given machine and operation.

$$N = \frac{a \times s \times ks \times Vc}{60 \times 102 \times E}$$

where N is the power kw (1 hp = 1.kw/0.745), a the depth of cut (mm), s the feed (min/rev) and Vc the cutting speed (mm/min). E is an empirical value for the efficiency of the machine, i.e. it takes into account losses in the drive which are mainly frictional. A machine with an enormous constant mesh gear box and feeds driven from the spindle would have a E of 0.5, while a machine with a simple gear box and, say, a d.c. motor which gives small losses would have E=0.8 or even greater if it had a separate feed motor. Ks is the specific cutting force for a given material. The value for Ks is obtained from Table 1.5 but must have one of the following deductions made: 5% for negative rake tips; 15% for positive rake tips; 10% for negative rake tips with a ground or sintered chip breaker.

Table 1.5

Material	Ks
Aluminium	70
Red bronze	75
Iron or mild steel up to 200 Brinell and malleable iron	110
Cast iron bars 200 to 250 Brinell	150
Cast iron bars 250 to 400 Brinell	175
Phosphor-bronze	175
Cast steel and bars from 50 to 70 kg/mm^2	210
Cast steel and bars from 70 to 100 kg/mm^2	230
Austenitic stainless steel	260
Tool steels from 100 to 180 kg/mm^2	300
Manganese steels (12% Mn)	360

CUTTING ANGLES

The cutting angles for high-speed steel single-point tools are given in Table 1.6 and those for cemented carbide in Table 1.7. With the latter no single grade is suitable for machining a range of different materials, and the maker's recommendation should be followed, for some firms supply up to eight grades. In general, cemented carbide tools are the type used in engineering workshops for turning operations, owing to the higher cutting speeds that can be used. There are still applications for high-speed steel, particularly for heavy roughing cuts on forgings, and also for tools such as twist drills, milling cutters and broaches. The cutting speeds for high-speed steel tools are in the following ranges: mild steel, 170 ft/min (50 m/min); cast iron, 120 ft/min (36 m/min); phosphor-bronze 100 ft/min (30 m/min); high-carbon steel, 35 ft/min (11 m/min); 100 ton nickel-chrome steel, 25 ft/min (8 m/min).

Cutting speeds for use with cemented carbide tools can be at least double those for high-speed steel, and higher still for ceramic tools (see Table 1.2).

Table 1.6
Cutting angles for high-speed steel tools

Material	True top rake (deg)	Clearance (deg)
Mild steel, 32 tons/in^2 tensile	20	6
Steel above, 50 tons/in^2 tensile	15	6
Cast iron	10	8
Phosphor-bronze	6	10
Brass	0 to 6	10
Aluminium	30	10

Table 1.7
Cutting angles for cemented carbide tools (positive rake)

Material	Top rake (deg)	Front clearance (deg)	Side clearance (deg)
Mild steel, 32 tons/in^2 tensile	10	5	7
Steel above, 50 tons/in^2 tensile	8	5	7
Soft grey cast iron	8	5	5
Phosphor-bronze	3	5	5
Yellow brass	3	5	5
Aluminium	18	7	9
Duralumin	13	7	9
Zinc-based alloys	13	7	9

TOOL LIFE AND CUTTING SPEED

The shape of a typical curve relating to tool life and cutting speed is shown in Figure 1.20. The curve has an equation of the type $VM^b = A$, where V is the cutting speed, M is the tool life to failure and b and A are empirical constants. These curves are fundamental to a study of the economics of machining operations. If the cutting speed for an operation is raised, the machining costs decrease, but, on the other hand, the tool life decreases also, resulting in increased grinding and tool costs. The most economical cutting speed for an operation is therefore the speed at which the sum of the machining costs and the regrinding and tool costs is a minimum. This indicates a compromise, and the most economical conditions are not, therefore, the conditions which give the longest tool life obtainable.

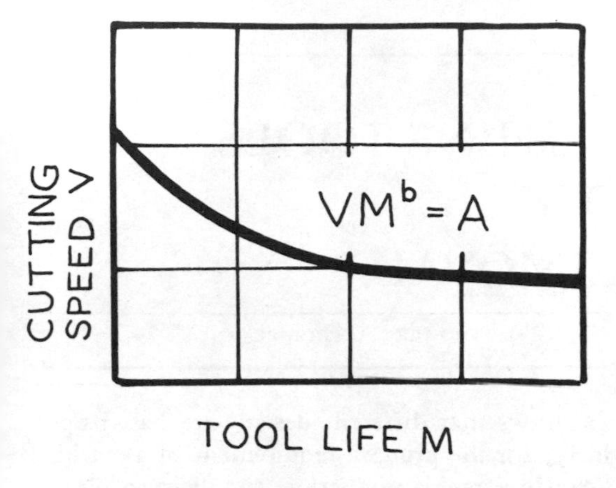

Figure 1.20.
A curve showing the relation between tool life and cutting speed

The cutting speed which will result in any desired tool life can be obtained from the curve, although this will only apply to the conditions for which the curve has been drawn. This may refer to a permissible cutting speed for a tool life of, say, 60 min. The tool life is the time of actual cutting, which represents a much longer apparent life on the machine. The permissible cutting speed for a stated tool life is also a useful measure of the durability of a tool under different conditions.

The values of the constant b in the equation which are observed in practice make the permissible cutting speed a much more sensitive measure of cutting performance than is the tool life between regrinds. An improvement in cutting material may give an increase in tool life at the same cutting speed or may permit an increase in the cutting speed for the original tool life. The percentage increase in tool life is substantially greater than the corresponding increase in cutting speed. For example, an increase amounting to 100% in tool life for the same speed is equivalent to an increase of only 9% in cutting speed for the original life. Improved results are thus quoted in terms of tool life, but whether improvements in tooling should be utilized in raising the cutting speed or in allowing the tool life to increase can be correctly decided only by consideration of the economics of the operation.

Tool regrinding is kept to a minimum by the use of replaceable tips on tools. The very process of regrinding tools should be restricted to those which must inevitably be special, i.e. those whose geometry is dictated by the shape of the workpiece which cannot be bought as a 'throw-away' tip. A tool- and cutting-grinding machine, and a man to operate it, represents non-productive overheads which add to the cost of each part made. Equally, a productive machine is losing money when it is stopped for a complete tool change (instead of a tip) and the necessary resetting of that tool.

2

TYPES OF MACHINE TOOLS AND METHODS OF METAL REMOVAL

A study of machine tools shows that different designs are based upon factors that often vary widely, for the primary requirement of a machine tool may be accurate production, rapid production, simplicity of design or adaptation to a wide range of work. It is because of these varying or related factors that machine tools, even of the same general type, are made in a number of different designs. Two factors which affect the usefulness of any machine tool are the ease of control and the relative position of the constituent parts. Mechanical, electrical, electronic, hydraulic or pneumatic systems may be used, and, in general, combined services give the best result.

The two main methods of metal removal are the machining of plane surfaces and the formation of cylindrical components, either external or internal. By means of simple diagrams, the main features of the various types of machine tools will be considered, commencing with reciprocating machine tools, i.e. shaping, slotting and planing machines; all of these use single-point tools to remove metal.

SHAPING MACHINES

The shaper is an inexpensive machine requiring little skill to operate. The cost of cutting tools is by comparison with, say, milling cutters very small. The general construction is shown in Figure 2.1(a) and comprises the ram A, which reciprocates forwards and backwards over the work so that the tool, which cuts on the forward stroke only, can produce a flat or angular surface. By tracer-copying devices various additional contours can be produced. To save time, the non-cutting return stroke is made at a faster rate than the cutting stroke, at a ratio of about 2 to 1.

The machine is started or stopped by lever B, and the speed is changed by the two levers C. The saddle mounted on slides on the front of the body can be elevated or lowered by hand or power at a variable feed rate by levers D. The motion then continues to gearing on the end of the cross slide to the vertical screw or to a horizontal screw in the cross slide to give the horizontal table feed motion. The length of the ram stroke can be varied from zero to the maximum by a lever on the squared shaft E, while

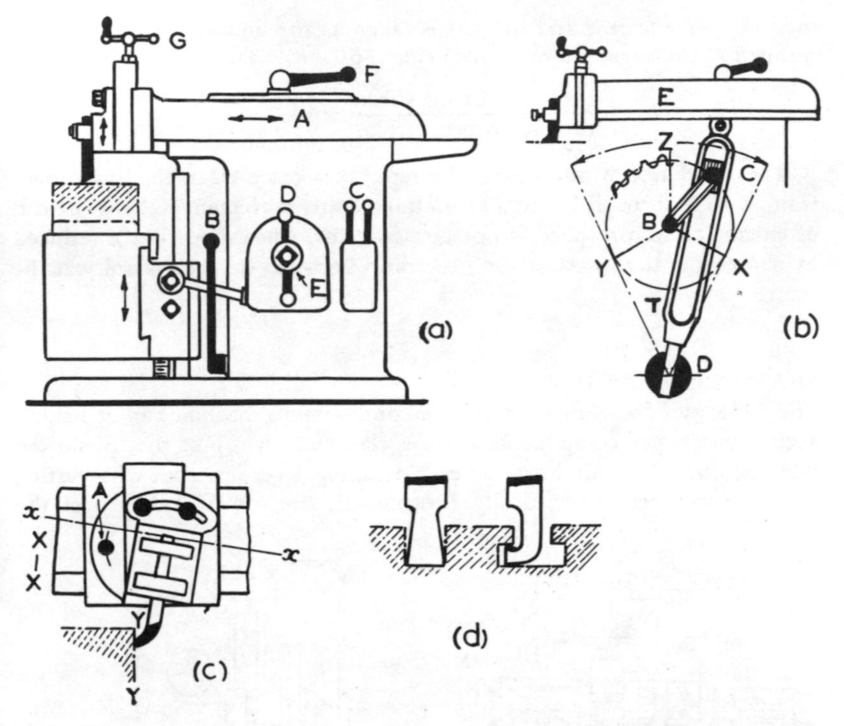

Figure 2.1.
A diagram of a shaping machine showing slotted-link motion and machining operations.
(a) The General construction:
A ram B, C and D levers E squared shaft G handle F lever
(b) The mechanism used for the tool traverse:
BC crank D trunnion E ram T slotted link
(c) Machining the vertical edge of a casting:
A bolts
(d) Tools used for cutting deep grooves or tee slots

the ram can be clamped in various positions in its slide by the lever F. The tool box can swivel through a wide angle, and the depth of cut can be adjusted by the handle G.

SLOTTED LINK MOTION

Figure 2.1(b) shows the mechanism generally used for the tool traverse. There is a crank BC adjustable for length and fastened to a wheel revolving around point B at a uniform speed. By means of a die that slides and drives the slotted link T, which is held but free to slide through a trunnion at D, the link is caused to swing and move the ram E backwards and forwards in its slideways. The ram is in its extreme position when the crank lies on the lines BY and BX, and the cut is taken as the link moves from X to Y through Z, the angular movement being $(360 - \theta)^\circ$. Thus

$$\frac{\text{time for cutting stroke}}{\text{time for return stroke}} = \frac{360 - \theta}{\theta}$$

The rapid return motion of the ram takes place when the link moves from Y to X through T, but a limitation of this movement is that the ratio of cutting to return speed is not constant, for, when the stroke is reduced by shortening the length of BC, the ratio becomes less and is unity at the centre.

SHAPING OPERATIONS

The majority of workpieces machined on a shaping machine can be held in a vice, but larger components can be clamped on top of the tee-slotted table or down the side of it. When machining a vertical edge of a casting with the tool feeding vertically downwards, the tool will drag over the

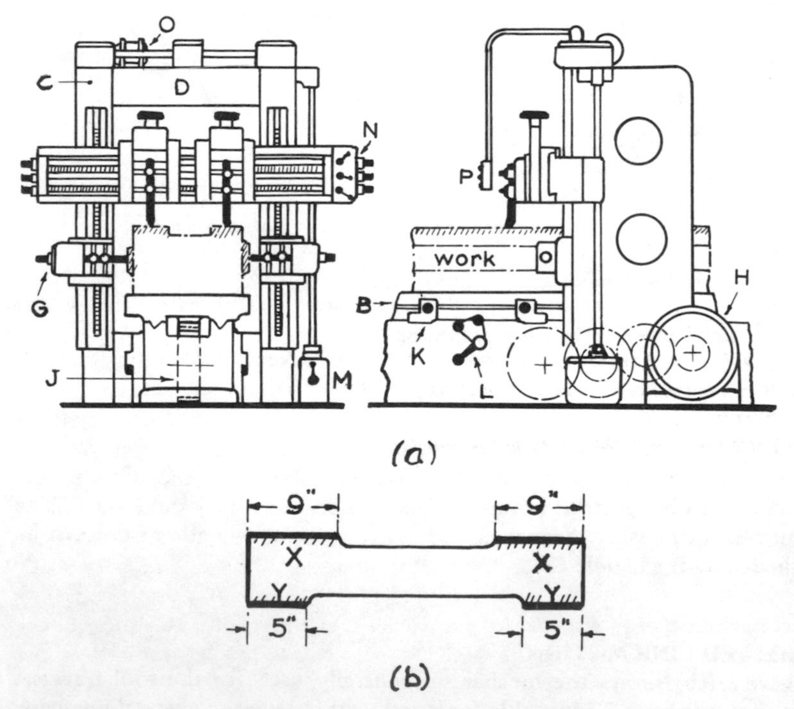

Figure 2.2.
(a) A diagram of a planing machine showing machining operation:
B table D cross rail J bull wheel M and N boxes
C uprights H motor K dogs O feed motor
(b) An example of machining the baseplate and time estimating

finished surface on the return stroke unless the tool box is set at an angle. The can be seen from Figure 2.1(c), for if the tool were set vertically it would swing upwards along the line YY, but, as shown, with the plane XX at right angles to xx, the axis of the swivel bolt, the tool point moves away from the work as it swings upwards.

When machining angular faces, it is necessary to loosen the bolts A, to swivel the complete tool slide to the angle required and then to prevent the tool dragging over the work on the return stroke; the tool should be further off-set as described. The rule is that the tool box should always be turned away from the surface that is being shaped or planed on vertical or angular faces.

Figure 2.1(d) shows the tools used for cutting deep grooves or tee slots. For tee slots, two cranked tools, left and right hand, are required. The operation is delicate, and tools may easily be broken, so that it is essential that the tool must be lifted clear of the slot during the return stroke.

SLOTTING MACHINES

The design follows closely that of the shaping machine but uses a vertical cutting motion. It is usually provided with a circular work table which can be traversed towards or at right angles to the tool and can also be rotated. This is because much of the work performed comprises the slotting of keyways or splines which require indexing in relation to each other. The Whitworth quick-return motion is often used in preference to the slotted link, for the Whitworth motion gives a constant ratio between cutting and return strokes, irrespective of the length of the stroke.

PLANING MACHINES

For the machining of large components it is preferable to mount the work on a travelling table which passes under the tools mounted on a cross slide. This unit can be elevated to accommodate various work heights, and the saddles carrying the tool boxes can be traversed along the cross slide to give the horizontal feed motion. A vertical motion to the tools is also available, while additional tool boxes with power feed motions can be fitted to the uprights. Figure 2.2(a) shows a typical machine operating on a bed casting. The table is driven from motor H through gearing which terminates with the bull wheel J driving the table rack. Slots on the table B accommodate dogs K which trip the lever L at the required distance of the table travel. The rate of the saddle feed motions can be selected on the box M, and the engagement can be made on the box N on the end of the cross slide. A feed motor O is mounted on the cross rail to elevate or lower the cross slide on the uprights C, which are of heavy box section tied at the top by the cross rail D.

Modern planing machines are reverting to a very old device, that of double cutting tool boxes to eliminate the idle return stroke, and cutting speeds to up 350 to 450 ft/min (106 to 140 m/min) are being used. Also, alternative systems of driving the table are used, e.g. the spiral pinion meshing with the table rack, or for shorter stroke machines hydraulic table traverse is used; this oil system is also used for some shaping machine ram drives.

PLANING OPERATIONS

It is is necessary to find the number of cycles (one forward and return stroke) of a definite length that a machine completes in a given time, then the speed in feet per minute (metres per minute) will be

$$\frac{\text{length of cycle (ft (m))} \times \text{no. of cycles} \times 60}{\text{time (s)}}$$

if we assume that a machine operates on a 10 ft (3 m) stroke, making 16 cycles in 4 min, then the average speed is

$$\frac{20 \times 16 \times 60}{4 \times 60} = 80 \text{ ft/min (24 m/min)}$$

To estimate the time to machine a baseplate 9 ft long as in Figure 2.2(b) using a feed of $\frac{1}{32}$ in/cycle for roughing and $\frac{1}{4}$ in for finishing cuts with an allowance for the table over-run and the tool cross traverse, we adopt the following procedure.

FIRST SETTING

The first setting is with the baseplate resting on face X. The first operation, is to plane face Y, with two tools cutting, and two roughing cuts only are made.

$$\frac{6 \times 2 \times 32}{4} = 96 \text{ min}$$

SECOND SETTING

This setting is with the baseplate resting on face Y. The first operation is to place face X, with two tools cutting and two roughing cuts.

$$\frac{10 \times 2 \times 32}{4} = 160 \text{ min}$$

The second operation is to finish planing face X, with two tools cutting, and one cut.

$$\frac{10 \times 1 \times 4}{4} = 10 \text{ min}$$

The actual machining time is therefore 266 min.

To obtain the floor-to-floor time we must add the following.

(1)	First setting and bolting on machine	30 min
(2)	Changing tools and grinding	30 min
(3)	Gauging	20 min
(4)	Taking work off machine, cleaning table	20 min
(5)	Second setting and bolting down	30 min
(6)	Allowance for contingencies and fatigue	30 min
	Total 266 + 160 =	426 min

GRINDING MACHINES

Two types of grinding machines for producing flat surfaces in Figure 2.3(a) and (b). For heavy metal removal a cup or ringwheel as in Figure 2.3(a) is used, sometimes with abrasive segments instead of a solid wheel. The spindle is driven by the vertical motor A and is provided with a traverse along the cross slide by shaft B and also vertical adjustment by the handwheel C. The table traverse along the bed, the rapid and fine cross traverse of the wheel head, the coolant pump motor and the starting and stopping of the table traverse and spindle motors are all controlled from the push-button station D.

Figure 2.3(b) shows the alternative method of surface grinding, by using a horizontal spindle to carry the grinding wheel which cuts on its

Figure 2.3.
Types of surface-grinding machines with vertical and horizontal spindles:
A vertical motor B shaft C handwheel D push-button station

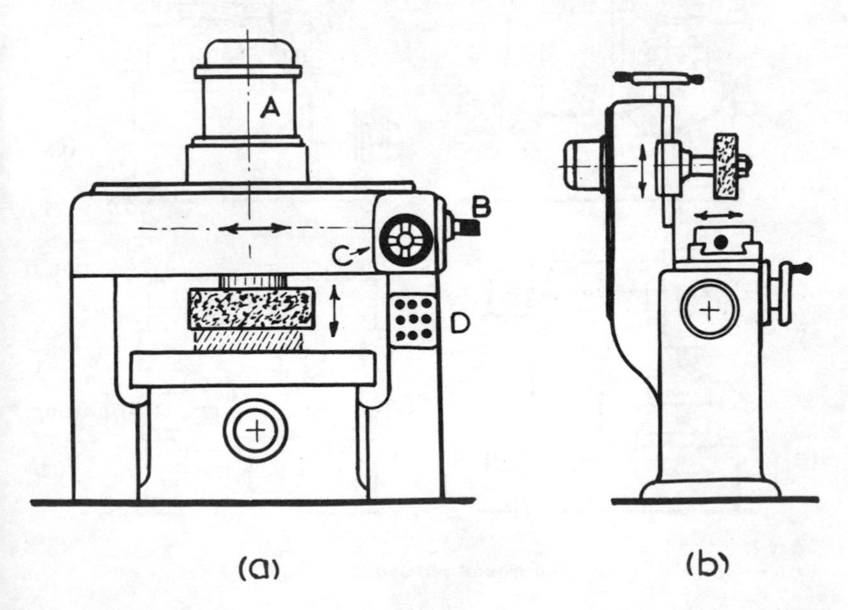

periphery. This is a tool room type of machine used for grinding small components such as flat gauges and parallels. There are three traverse motions, i.e. (1) a vertical traverse of the spindle to apply the depth of cut, (2) a variable speed of the table longitudinal traverse on the bed and (3) a cross feed motion of the table. Further details of the grinding process are given in Chapter 6.

MILLING MACHINES

The subject of milling operations is discussed in Chapter 5, but if we consider the construction of the machines, Figure 2.4 shows in diagram form the four principal types. Plain and universal machines (Figure 2.4(a)) are similar in construction, but the universal type has certain additional features such as a swivelling table and attachments which enable more intricate work such as spiral milling and indexing to be performed. The plain machine is more rigid in construction and thereby better adapted for ordinary workshop conditions. It has rectangular movements only to the table and is not fitted with dividing heads as regular equipment. The machines are of the Knee type, vertical adjustment being made by the table.

For plain manufacturing purposes it is preferable to dispense with a cantilever table and to support the work on a box bed as shown in Figure 2.4(b). The vertical adjustment is then obtained by elevating or lowering the saddle and support. As the machine is intended for mass production

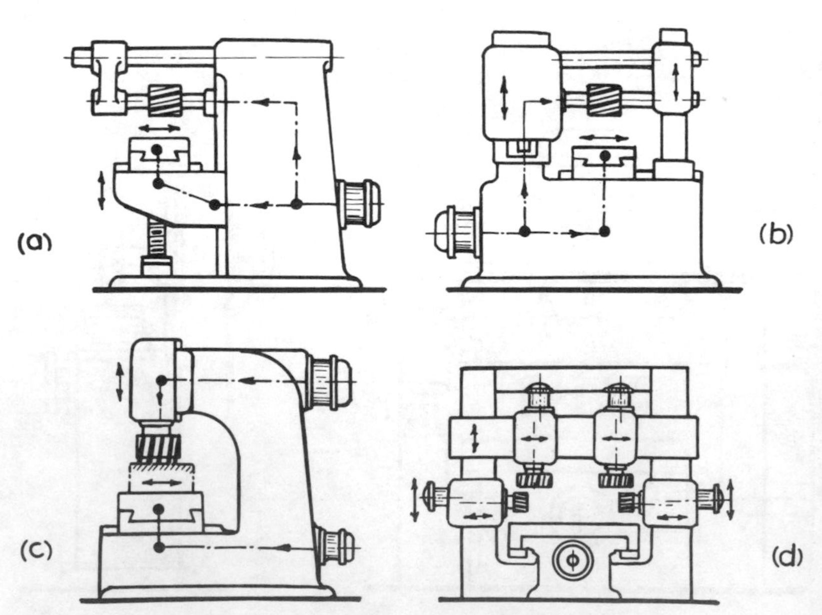

Figure 2.4.
Diagrams of various types of horizontal and vertical milling machines

purposes, only a limited speed and feed range is required, and the machine may often work on an automatic cycle.

The Vertical milling machine (Figure 2.4(c)) may have a table supported on a bed as shown or may be made with a knee-type table as in Figure 2.4(a). The main advantage for vertical milling is that the construction is favourable for work inspection during machining, while the rigid support of the cutter is favourable for heavy cutting. The usual operations are surface machining, edge milling, tee slots and recessing.

The plano-milling machine (Figure 2.4(d)) has come into much prominence during recent years and takes its name from its constructional similarity to a planing machine and its ability to perform similar work. The machine has an elevating cross slide to carry the milling saddles which may be duplicated on the uprights. A separate motor on each saddle simplifies the construction considerably.

LATHE DEVELOPMENTS

The lathe is the most common type of machine tool, and the construction of a standard lathe is fully described in Chapter 3. It is also the most versatile, for by suitable attachments its range of work can be extended so that it can often compete on batch- or mass-produced components. These attachments include the use of hydraulic or pneumatic chucks, or turret rests instead of the normal compound rest. In some cases where the tailstock is dispensed with, a saddle with a large hexagon turret can be employed, thus introducing the boring and surfacing lathe.

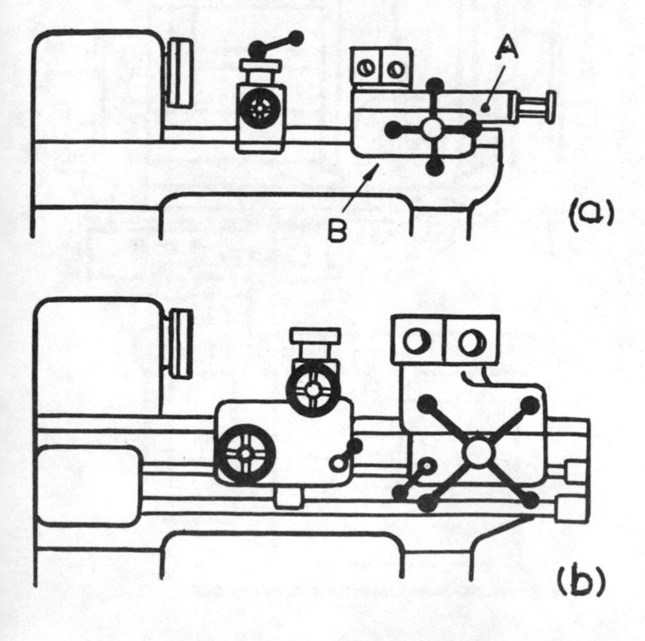

Figure 2.5.
Diagrams comparing (a) the capstan and (b) the turret lathes:
A slide
B saddle

The capstan lathe (Figure 2.5(a)) and turret lathe (Figure 2.5(b)) show how the development towards high production has occurred, for, by fitting two saddles, several tools can be in operation together, while others are readily available without having to change tools after each operation. To define the difference between the machines shown, Figure 2.5(a) shows that on a capstan lathe the turret is mounted on a slide A, with a longitudinal movement in a saddle B fixed to the bed, while a turret lathe (Figure 2.5(b)) carries its turret on a saddle which slides directly on the bed. Both types are usually fitted with an additional square turret, or cut-off rest, fitted on the bed close to the chuck. As indicated, if power traverses are available to the square turret, the machine becomes a combination turret lathe (Figure 2.5(b)). Capstan lathes normally operate on smaller components from bar than do turret lathes which machine large forgings or castings, although some turret lathes will accommodate a bar of 8 in (203 mm) through the hollow spindle.

BORING AND TURNING MILLS

Boring and turning mills may be regarded as vertical lathes operating on components such as pulleys and large gear blanks where the diameter is

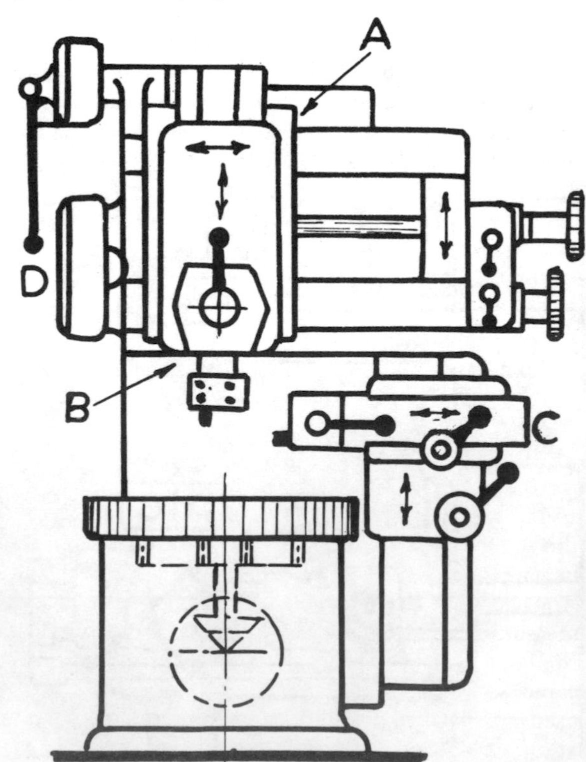

Figure 2.6.
A boring and turning mill of the vertical turret lathe type:
A saddle
B turret slide
C side head
D traverse lever

great in relation to the depth. The production can be high owing to several advantageous features in construction.
(1) The work setting is easy compared with mounting work on a vertical faceplate, for any component placed on the table can be clamped by the operator without difficulty.
(2) All the weight of the work and table is vertically downwards on a large spindle bearing.
(3) Balancing of the work is not as necessary as when mounted on a vertical face, for gravity does not come into consideration.

Figure 2.6 shows a type best described as vertical turret lathe. The saddle A has a longitudinal traverse on the cross slide which can be elevated or lowered on the uprights. The turret slide B provides the vertical feed for the tools in the pentagon turret, while other tools can be in operation from the side head C which has its own movements on the side upright. The cross slide can be elevated or lowered by lever D, while the other levers shown are for engaging the various traverse motions. Alternative designs are arranged with two saddles on the cross slide together with a side head, while other types have two work tables with separate drives and controls so that two components can be machined simultaneously. The largest machines may have a table 60 ft (19 m) in diameter.

MACHINING OF PUMP PLUNGERS

Figure 2.7 illustrates the advantages of multiple tooling. Figure 2.7(a) shows the tools used on a standard lathe with the bar held in a chuck. Eleven operations are required including the use of three tools in the compound rest and four in the tailstock. The various parts of the plunger are numbered with the same figures as the tools performing the operations.

(a) Turn diameter 7 full length.
(b) Turn diameter 4.
(c) Square out 5, 6 and face end 7.
(d) Cut shoulders 1, 2 and 3.
(e) Centre and recess end of bore 9 from tailstock.
(f) Drill main bore 10 with twist drill.
(g) Drill small bore 11 using extension socket.
(h) Ream main bore.
(i) Cut off to length using tool 8.

If now a capstan lathe is available, the same tools could be arranged on the square and hexagon turrets as shown in Figure 2.7(b). The main saving in time is because every tool is in position in the hexagon turret. In addition, stops are set to limit the forward traverse of the tools, so that depth measurement is not required. If now it is assumed that large quantities of the component are required, a more elaborate set-up can be used as indicated in Figure 2.7(c). The tool 7 is taken from the square turret and is used in conjunction with the drill 10, so that turning and drilling proceed at one traverse.

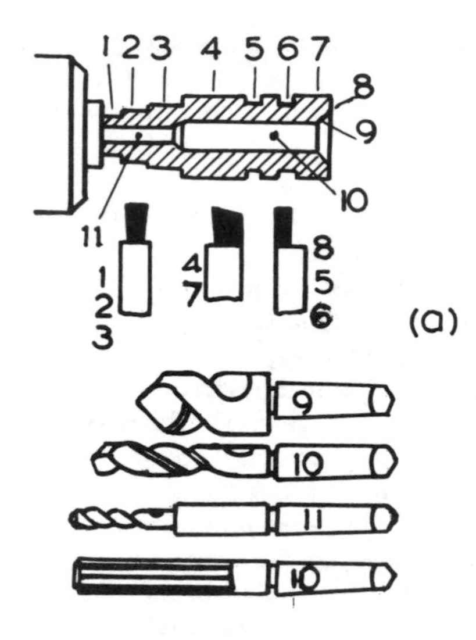

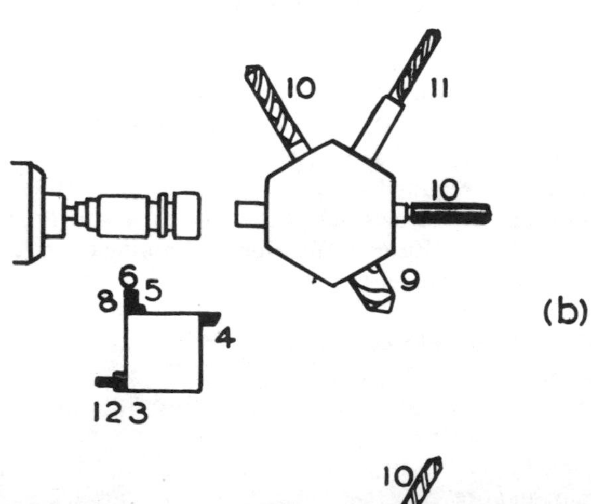

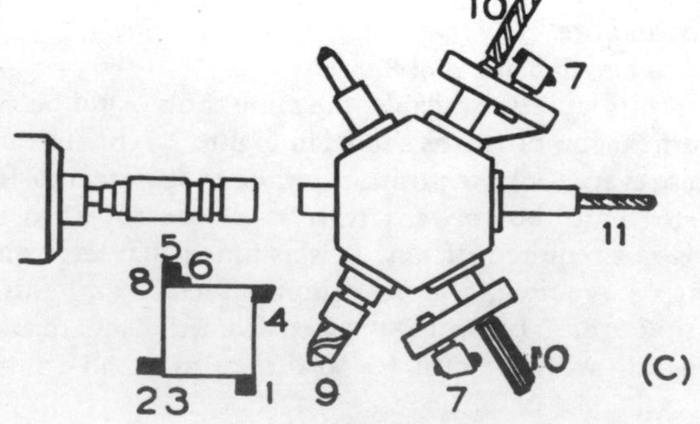

Figure 2.7. *Tooling operations for machining pump plunger. For explanation of the numbers see text*

A comparison of the three methods (a), (b) and (c) shows the following.

(a) The machining time, including trial cuts, moving tools and tailstock is 60 min per piece, i.e. 600 min for 10 pieces.

(b) Changing tools takes 15 min, adjusting tools to size 17 min and setting stops 13 min; this gives 45 min. The machining time is 25½ min, i.e. 255 min for 10 pieces; thus the total time is 300 min.

(c) This arrangement is for a batch of 40 pieces, and the machining time is 19 min per piece i.e. 760 min for 40 pieces. Adding 180 min for setting up gives 760 + 180/40 = 23½ min.

Thus the times are 60, 30 and 23½ min per piece for methods (a), (b) and (c) respectively.

DRILLING MACHINES

The operation of drilling is that of producing a hole in solid metal, whereas boring is the enlarging of an existing hole. The drilling machine was one of the first machines to be developed, for the production of holes forms a large part of all engineering operations, and the diversity of types is greater than that of any other machine tool. Designs differ from the simple machine with one spindle to huge transfer machines with unit heads; some of these heads may have as many as 50 spindles on one head.

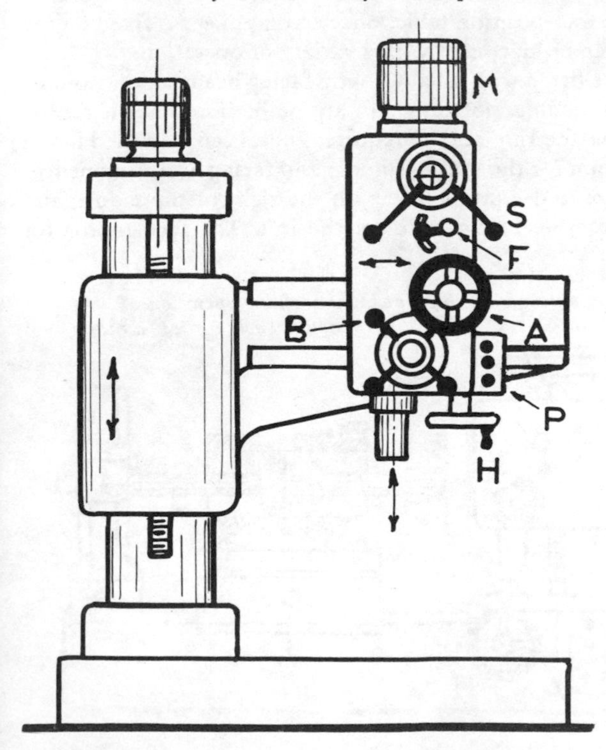

Figure 2.8.
A diagram of a radial drilling machine showing the control elements:
A handwheel
B star wheel
F lever
H handwheel
M motor
P push-button station
S levers

The radial drilling machine is a familier type of general-purpose machine and is shown in diagram form in Figure 2.8. For convenience of operation nearly all the control elements are mounted on the saddle, including the motor M for the spindle speeds which are changed by levers S, while the feed changes are by lever F. Hand traverse to the spindle is by the handwheel H and the traverse of the saddle along the arm by the handwheel A. The feed engagement is by the star wheel B.

The arm is positioned on the column by a nut and screw driven from the motor on top of the column, and starting and stopping of both motors is from the push-button station P on the side of the saddle.

BORING MACHINES

Figure 2.9 shows a horizontal boring machine; this type of machine is versatile in that not only boring and drilling operations can be carried out, but also facing operations can be carried out by a tool in the head F. Similarly, by using cutters either in the spindle nose or attached to the facing head, milling operations can be carried out by hand or power traverses. These operations are feasible because of the available feed traverses, i.e. the feed to the boring bar, the elevation of the saddle on the upright, the table longitudinal traverse and also the transverse traverse. Thus, by the use of a four-position table, once a component is fixed on the table, all four faces can be machined with a variety of operations.

Many machines are fitted with an automatic facing head; so, on work of the type shown on the table, not only can a pipe be bored but it can be faced and turned over the rim, and, if required, holes can be drilled in the rim. The driving motion for the boring spindle and facing head commences with motor M, with speed control levers on the right of the saddle, and feed control levers for the facing slide on the left. The feed motor for

Figure 2.9.
A diagram of a horizontal boring machine showing machining operations:
A box B feed motor F head H handwheel M motor

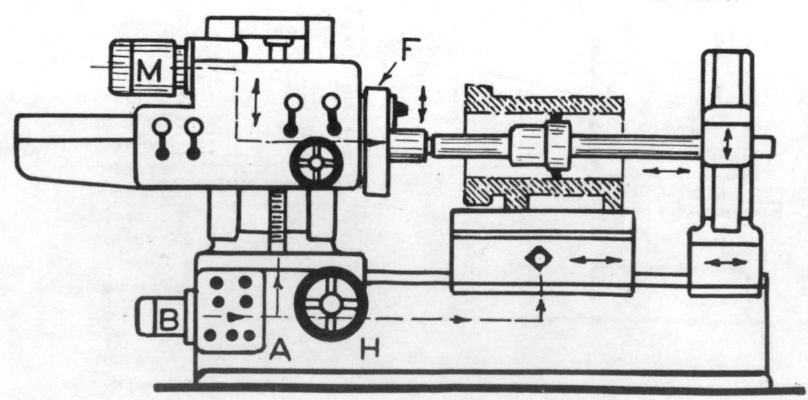

elevating the saddle and the table traverses on and across the bed is shown at B. The directional traverses are selected by handwheel H, while the feed rates are changed in the box A which also carries push buttons for motor control.

CYLINDRICAL GRINDING MACHINES

Following turning operations vast quantities of shafts and spindles are brought to size on a cylindrical grinding machine shown in diagram form in Figure 2.10. Three separate motor drives are required, one shown at A for rotating the workpiece, a second at B for rotating the grinding wheel and a third motor C for the table traverse usually by hydraulic power from a pump and cylinder. Stepless table speeds are thus obtained as selected on the dial D, while four speeds usually suffice for the work head; these four speeds are provided by a four-speed motor controlled by the handwheel E. The table hand traverse is by the handwheel F, the length of the traverse being regulated by the adjustable dogs contacting the reverse lever. The power traverse is by the lever G. The wheel in-feed is by the handwheel H for the hand motion and by the lever J for the power traverse. Push-button control is provided for the motors.

INTERNAL GRINDING MACHINES

Internal grinding machines (Figure 2.11) are employed for the sizing of bores of gears, ball races and other close-fitting components. Unlike the cylindrical grinding machine where the grinding wheel can be of a large diameter, in internal grinding the wheel diameter is limited by the size of

Figure 2.10.
A cylindrical grinding machine showing the control elements:
A, B and C motors *D dial* *E and F handwheels* *G lever*
H handwheel *J lever*

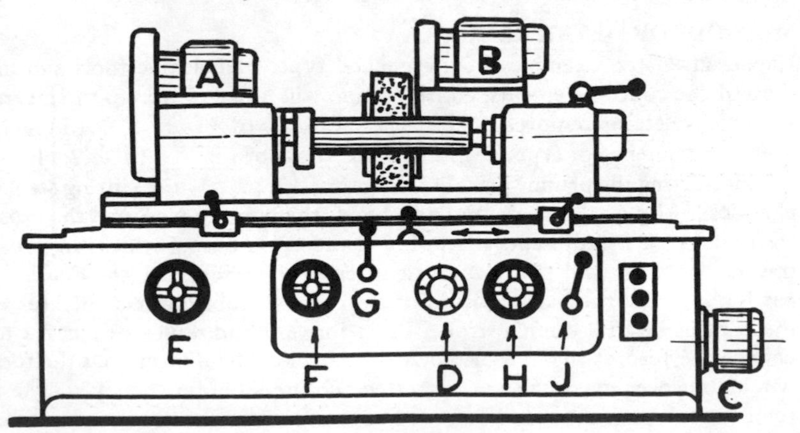

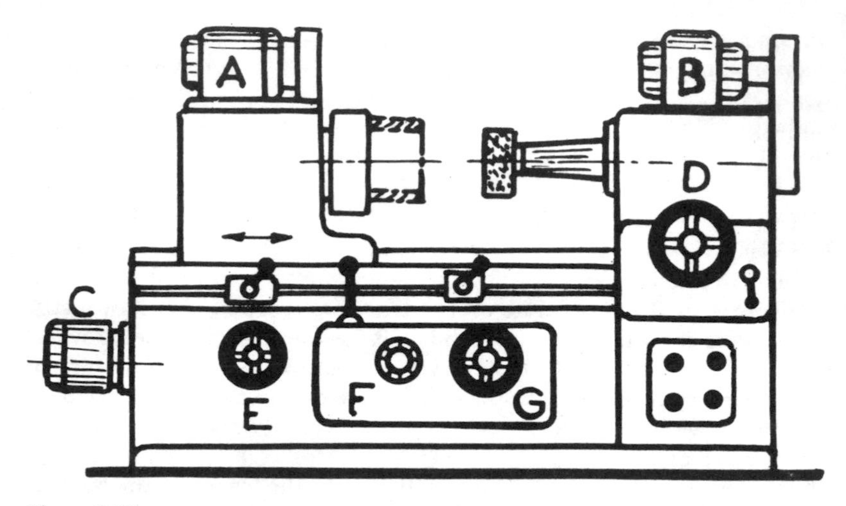

Figure 2.11.
A diagram of an internal grinder showing motor drives:
A, B and C motors D and E handwheels F dial G handwheel

the bore, so that wheel wear is generally rapid. Again three main motors are used, A for revolving the work, B for revolving the grinding wheel and C for the workhead travel along the bed. The handwheel D is for the hand cross feed to adjust the depth of cut, while the adjacent lever engages the power feed. The handwheel E controls the speed of the workpiece motor, and the dial F controls the speed of the hydraulic workhead traverse on the bed. The handwheel G is for the hand traverse. The length of the workhead traverse is set by the adjustable dogs which contact the reverse traverse lever on top of the control box. The starting and stopping of the various motors is by the push buttons shown.

METHODS OF METAL REMOVAL

Having given the examples of the general types of machine tools and indicated the control elements, consideration will now be given to the means by which metal is removed. The parts (1) to (8) of Figure 2.12 follow in the same sequence of types of machines as those of Figures 2.1 to 2.11.

Considering the planing machine, (Figure 2.12, part 1) the cutting stroke is indicated by *C*, the tool approach by *A*, the overrun by *B* and the cross traverse by *D*. *F* represents the point in the cycle during which the cross traverse operates and *G* the amount of cross traverse, while the depth of cut is shown by the black portion of the cut. The tool traverse operates at the conclusion of a double stroke, i.e. a forward and return stroke, for to change the feed at the end of the cutting stroke would mean that the tool would drag over the rough casting and would probably be destroyed. These remarks only apply to machines without automatic lifting of the tool box on the return stroke, and this can be considered an essential feature when planing with cemented carbide tools. Similarly, when planing by using a double-cutting tool box, the feed would be applied at the beginning of the normal cutting stroke, for with double cutting the return cut is often less than the previous one to give a fine work finish.

The shaping machine feed on the type of the machine in which the traverse is actuated by crank and ratchet motion is applied practically throughout the whole of the stroke. It is better, however, if the cut can be applied practically instantaneously by cam action, as in the slotting machine, and in a constant time relation to the length of the stroke.

SURFACE GRINDING

In Figure 2.12, part 2, *W* is the length of work, *L* the total wheel travel, *A* the necessary overrun for the feed to operate, *C* the depth of cut and *T* the transverse of longitudinal feed. For light duty, a horizontal spindle mounting of a peripheral wheel may be used to traverse from end to end

Figure 2.12.
Diagrams showing metal removal methods on various machine tools

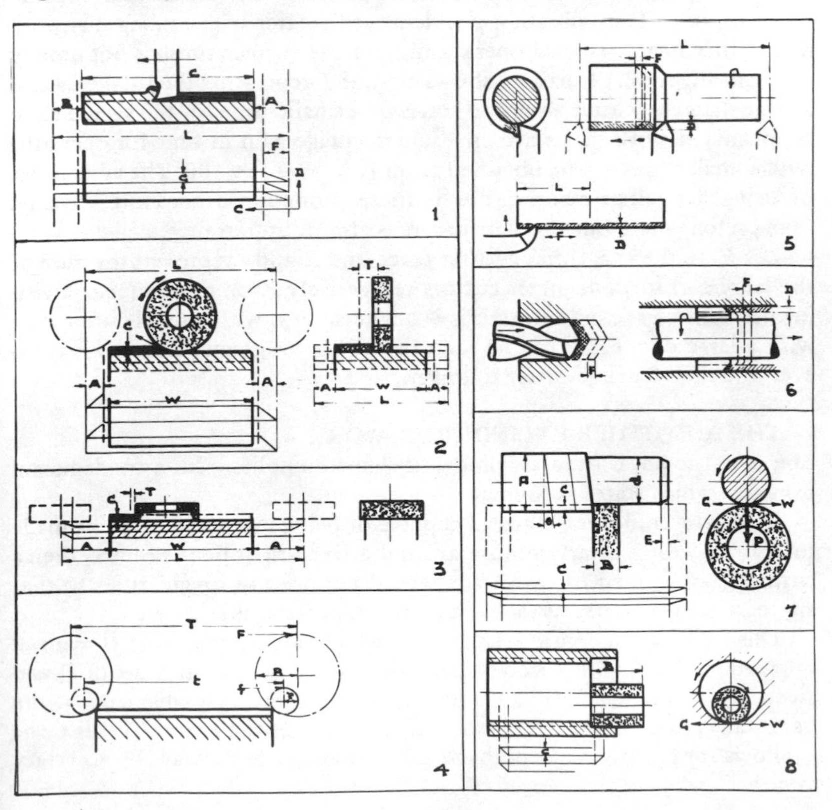

of the work and to feed across about four-fifths of its width per stroke. Again the time taken to complete the machining is largely dependent upon wheel width.

Figure 2.12, part 3, shows a better method for heavy duty. The cup-shaped wheel can cover a large work area, sufficient in many cases to dispense with any cross traverse, and this coupled with the vertical spindle mounting which is not detrimental to spindle bearings as a horizontal wheel, allows heavier cutting to take place. The longitudinal feed must be finer than in the first case owing to the amount of wheel in contact, if the same power is employed for the drive, and the overrun *A* is also greater as shown. The two methods produce surfaces with entirely different lubricating characteristics.

MILLING OPERATIONS

The operation of milling is based upon the rotation and feed of a multi-tooth cutter. The metal removal is shared by several teeth at each revolution of the cutter. With a single-point tool in operation, the cutting forces remain steady, if not constant, but with a milling cutter in action the forces vary during the cut as each tooth strikes the work and produces a trochoid shape of chip. Thus vibration problems are greater with milling than with most other metal removal operations, and the surface finish is not usually as good. Figure 2.12, part 4, shows that, if T represents the total travel of a large diameter cutter and t the travel of a smaller one for the same length of machined work surface, then the percentage gain in time for operating with a small cutter can be obtained from $[(T - t)/T] \times 100$. The advantages of using a small-diameter cutter is more pronounced on short traverses than on long ones, but another feature is also of importance.

Let F be the tangential cutting force and R and r represent the radii of the large and small-diameter cutters respectively; then the torque or twisting moment on the machine arbor is obviously less with a small cutter than with a large one, for it can be seen that $F \times R$ is greater than $f \times r$. However, the life of a large cutter is longer.

LATHE AND OTHER CYLINDRICAL WORK

The metal removal from a cylindrical shape complies with a fundamental principle which states as follows.

By whatever means a material capable of being machined could be made to revolve with a rotary motion around a fixed right line as centre, then a cutting tool applied to its surface would remove the irregularities so that any part of such surface should be equidistant from that centre.

This is shown in Figure 2.12, part 5, which demonstrates metal removal by turning. The work revolves, and the tool being set at a depth D and given a longitudinal feeding movement F will reduce the diameter of the bar by an amount $2D$ as it moves in the direction of the arrow. The distance L allows for the overrun at both ends, but this may not always be necessary at either end.

The speed of cutting will vary with the reduction in diameter of the bar but not to the extent of a facing operation shown in the lower diagram. Here it can be seen that the surface speed of rotation is zero at the centre and increases with the diameter of the work, so that a tool travelling outwards from the centre is cutting at a different rate throughout its travel *F*. To overcome this drawback the speed of work revolution should automatically increase as the tool approaches the centre, for the feed is based upon inches per revolution (millimetres per revolution): the greater the revolution, the quicker the operation is completed.

Consider a disc 24 in (610 mm) in diameter with a 4 in (102 mm) bore. The surfacing traverse of 10 in (254 mm) in length with a constant cutting speed of 320 ft/min (97 m/min) and a feed of $\frac{1}{64}$ in/rev (0.04 mm/rev), the rotational frequency will vary from 58 rev/min on the large diameter to 320 rev/min when the tool reaches the bore. This gives an average of 189 rev/min or 3.5 min for the traverse.

Assume now that the cut commences with a speed of 320 ft/min (97 m/min) on the outside edge and that without altering the cutting speed it completes the traverse with the same feed rate. The rotational frequency will be 58 rev/min throughout, and the time per traverse will be 11 min or over three times as long as the previous case. The operator may make one or two speed changes, but the actual truth is that if it means stopping the lathe, as it often does to make a speed change, he may be inclined to leave the speed as first selected. Moreover, scrolling of the work usually takes place by starting and stopping in the middle of a cut. For driving purposes stepless speed rates are available by using d.c. motors with thyristor control, while for stepless feed rates hydraulic control is prominent.

DRILLING AND BORING

The operation of drilling (Figure 2.12, part 6) may be defined as that of removing metal from the solid by rotating and traversing a tool to produce a hole. The rotational speed permissible depends upon the outside diameter of the drill, the material of the drill and the material of the workpiece, while the rate of feed *F* depends upon the cuts per inch (cuts per millimetre) of the traverse. This is often given as a fraction of an inch (millimetre) per revolution of the drill.

Boring is the operation that generally follows drilling to enlarge the hole or to machine a cored hole in a casting. A single-point tool may be used for finishing, but for heavier metal removal a double cutter as shown may be employed. *F* again indicates the feed and *D* the depth of metal removed.

CYLINDRICAL GRINDING

It will be seen that in cylindrical grinding (Figure 2.12, part 7) the direction of wheel and work is such that they oppose one another to give a cutting action, for otherwise the wheel would merely drive the work. The peripheral wheel speed is about 7500 ft/min (227 m/min) as against a work speed of, say, 60 ft/min (18.3 m/min); this indicates the excess of pressure in direction *G* as against *W*.

The longitudinal feed is shown operating from right to left, and to prevent scrolling of the work the feed per revolution of the work must be less than the wheel width *B*. The in-feed can be applied at either end or both ends of the table traverse, resulting in a pressure *P* on the wheel head bearings. The wheel should never leave the ends of the work, or a result as shown at *E* is obtained.

INTERNAL GRINDING

The rate of metal removal on an internal grinding machine (Figure 2.12, part 8) is unavoidably much less than on a cylindrical machine, because the grinding wheel is small and must therefore be mounted on a small-diameter spindle. The overhanging position of the wheel and the small bearings of the spindle also tend towards a loss of rigidity. Wheel wear is rapid owing to the large area of contact between the wheel and work, and the larger the diameter of the wheel in relation to the hole to be found, the greater is the tendency to heat the work; this is because internal grinding wheels do not have the volume to dissipate the heat generated that external wheels do possess.

It is very important that the grinding wheel should not leave the bore during the grinding process, for the spring in the small-diameter spindle will cause the work bore to be bell mouthed. To prevent this happening,

Figure 2.13.
A combination turret lathe showing hexagon and square turrets with pre-selection of ten spindle speeds and ten feeds

not more than one-third of the wheel width should leave the bore during grinding. A wheel diameter of two-thirds of that of the hole is generally suitable for small work.

This chapter has been concerned with descriptions and diagrams of standard machine tools to indicate the general appearance and metal removal systems of each type. At the end of this and also some of the other chapters developments of the more advanced machine tools will be discussed in order to indicate the specialized design elements that high production demands.

Figure 2.13 shows a combination turret lathe with pre-selection of the spindle speeds and feeds, each of these having ten changes. Push-button

Figure 2.14.
A manually operated vertical boring and turning mill with a side head to increase the machining capability. The sidehead incorporates a turret, a vertical slide with a turret and electro-magnetic copying and a left-hand ram which may turn or grind
(By courtesy of Flli Morando & Co., Turin)

control is used extensively on the headstock, and an automatic version of the machine, equipped with tools on the hexagon and square turrets, is shown in Chapter 8.

Figure 2.14 shows a vertical boring and turning mill (Flli Morando & Co.) with a side head incorporating a turret. A vertical right-hand slide carries a second turret, while the left-hand ram may be used for turning or can be fitted with a grinding wheel for finishing a component by means of a wheel on the built-in spindle. Electro-magnetic copying is fitted, and the machine may be controlled by the operator during machining or when setting up work on the table, by means of the pendant control. This can be swung around to the most convenient position for the operator. The large-diameter spindle bearing which can be fitted on this type of machine, makes for high metal removal by the simultaneous operation of three tooling stations on certain types of work.

Amongst the large machine tools, the horizontal boring machine (Figure 2.15) (Mitsubishi Heavy Industries) is prominent, for in every branch of the engineering industry the largest structural members require boring and facing operations often spaced over a considerable distance. Thus a machine such as that in Figure 2.15, which is shown operating on large stands for rolling mills, requires a large column for the saddle vertical traverse. A platform is provided for the operator to travel with the machining head, and a feature of these machines is the limitation of manual levers which are superseded by pendant control for all spindle speeds and saddle motions, as well as for the travel of the column on the bed.

Figure 2.15.
A large horizontal boring machine operating on stands for steel rolling mills. The operator travels with the machining head
(By courtesy of Mitsubishi Heavy Industries, Japan)

3

THE LATHE AND MACHINING OPERATIONS

CONSTRUCTION AND OPERATION OF THE LATHE

The lathe ranks as one of the most important machine tools owing to the number of cylindrical components which are required for shafts, gears and other transmission units. The usefulness of the lathe is not restricted to this type of work, for a great amount of surface work can be machined by facing operations, whilst the facility for cutting screws is another indication of the versatility of the machine.

There is an immense diversity in design, ranging from the simple bench lathe to massive machines several hundred tons in weight. Again, lathes for specialized operations widen the range considerably, while machines equipped with multiple tools for mass production purposes are coming into more prominence. All these, however, are developments of the simple centre lathe which was introduced by Henry Maudslay in the year 1800.

UNITS OF CONSTRUCTION

Figure 3.1 shows the Harrison lathe with a capacity of 400 mm swing. The main units are indicated by reference letters.

BED B

This unit provides location for the headstock at one end and the tail-stock at the other. Induction-hardened and ground slideways on the top surface provide the guidance for the saddle and tailstock movement.

HEADSTOCK H

18 spindle speeds are available for rotating the work, these ranging from 40 to 2000 rev/min, the changes being by hardened and ground sliding gears. Speed selection is made through the dial on the centre of the head-stock. Other units include a push-button station with controls for the main motor, brake, emergency stop and coolant pump.

TAILSTOCK T

This locates on separate guides on the bed and can be clamped in any convenient position to accommodate the length of work. The two levers

Figure 3.1.
A 400 mm swing lathe showing the ergonomic controls:
A apron B bed H headstock S saddle T tailstock
(By courtesy of T.S. Harrison & Son Ltd, Hecmondwike)

shown are for clamping the tailstock to the bed and the spindle in the barrel.

SADDLE S
This unit rests on the bed slideways and carries the compound rest with hand and power traverses across the bed for the surfacing motion. Hand and power traverses are also available for the sliding motion along the bed.

APRON A
This is connected to the saddle and carries the mechanism for engaging the sliding, surfacing and screw-cutting motions; these are interlocked so that conflicting motions cannot take place. On the bottom shaft of the apron is a stop and start and forward and reverse lever.

FEED GEAR BOX F
This unit is driven from the end of the headstock and provides a range of feeds for the tool or alternatively through the lead screw for cutting a universal range of British or metric threads.

A feature of the design of the lathe is the ergonomic principle incorporated in the grouping of the control levers on the headstock, the feed gear box and the apron. All speeds are arranged in a geometrical progression, the usual speed ratio for machine tools being 1.26 to 1, i.e. if the first speed in a range was 30 rev/min, the next would be 30 × 1.26 = 37.8 and so on; the ratio r is found from

$$\log r = \log a - \log \left(\frac{b}{n-1}\right) \text{ or } - \log b/(n-1)$$

where a is the highest and b the lowest speed and n is the number of speeds in the range.

LATHE TRANSMISSION

If we consider a smaller lathe for simplicity, Figure 3.2 shows the transmission elements that effect the various motions. The headstock comprises sliding gears to give eight speeds from 64 to 1,440 rev/min. The gears on the first shaft slide in turn to engage the four inner gears on the intermediate shaft, while the two gears on the spindle slide in turn to engage the two outer gears on the intermediate shaft.

With the driving shaft running at 1,500 rev/min, the fastest speed is obtained by engaging the following gears

$$\frac{1500 \times 45 \times 56}{1 \times 46 \times 57} = 1{,}440 \text{ rev/min}$$

Figure 3.2.
A diagram showing the transmission elements of a lathe: A, B and C change wheels D knob

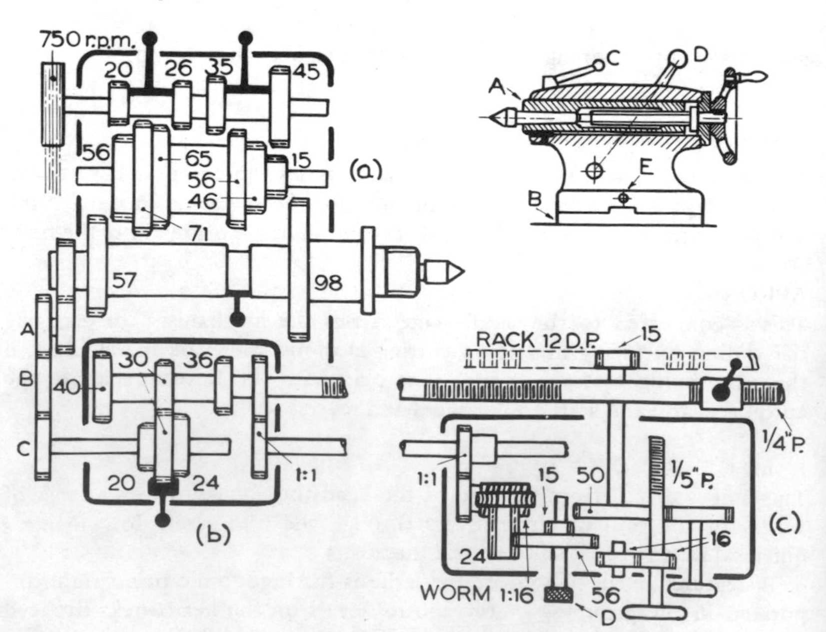

The slowest is given by

$$\frac{1{,}500 \times 20 \times 15}{1 \quad \times 71 \times 98} = 64 \text{ rev/min}$$

From the headstock, change wheels A, B and C connect to the feed gear box, giving three rates of traverse by sliding the gears on the bottom shaft. The transmission then continues to the apron through a pair of equal spur wheels to a worm and wheel reduction. The selection of either the sliding or the surfacing motion is made by moving the knob D, so that the pinion of 15 teeth engages with the 50-tooth wheel on the rack pinion shaft for sliding, or the wheel with 56 teeth engages with the pinion 16 teeth to connect up to the surfacing screw.

With the wheels in the feed box engaged as shown, the middle of the three feeds will be for sliding

$$\frac{30 \times 1 \times 24 \times 15}{30 \times 16 \times 56 \times 50} \times \left(\frac{15}{12} \times \frac{22}{7} \text{ rack pinion}\right) = 0.031 \text{ in/rev of spindle}$$

and for surfacing

$$\frac{30 \times \quad 1 \times 24 \times 56 \times 16}{30 \times 16 \times 56 \times 16 \times 16} \times \left(\frac{1}{5} \text{ surfacing screw}\right) = 0.018 \text{ in/rev of spindle}$$

LATHE CUTTING TOOLS

Figure 3.3 shows a representative set of lathe tools to cover machining operations encountered in the workshop. With lathe tools the cutting edges cover about 20% of the overall length, while the remainder is required for clamping in the tool post. Owing to the comparative high cost of high-speed steel, efforts are made to reduce the quantity required for the cutting portion, and a cheaper material is used for the shank; one method of achieving this is by butt welding the cutting portion to a shank of 0.5 carbon steel.

Another economical method is the use of tool bits which fit into holders as in Figure 3.4, although about one-third of the tool is wasted because of the length required for holding. The holder A is a three-way adjustable holder, B a straight holder, C a parting or cutting-off tool and D a type of

Figure 3.3.
A set of lathe tools to cover a range of machining operations

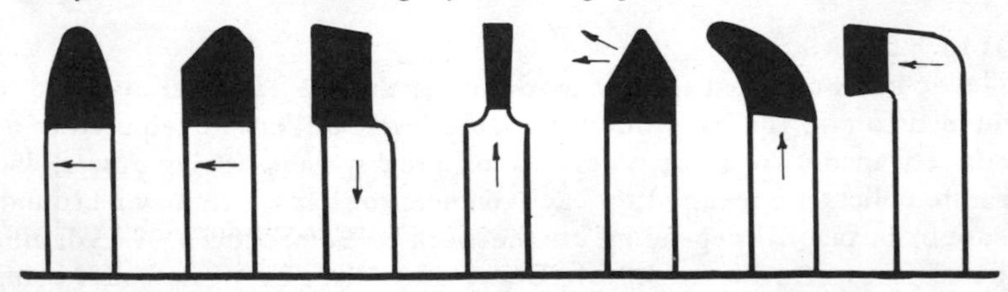

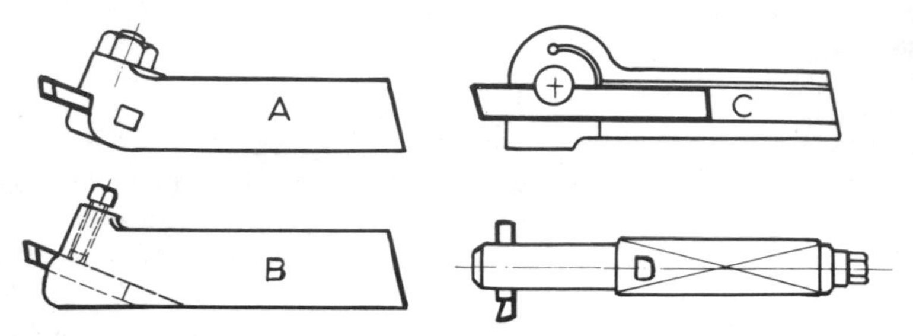

Figure 3.4.
Holders for turning, parting and boring tool bits

boring tool. With the more expensive and more brittle cutting materials such as carbides and ceramics, these are invariably used in the form of tips brazed onto steel shanks or as 'throw-away' tips clamped on a holder so that various edges can be used in turn.

WORK-HOLDING EQUIPMENT

Workpieces can be mounted between the lathe centres or can be held in a variety of chucks. Centre work usually applies to long work such as shafts which require support at each end, while short work of the flywheel or gear blank type can be held in a chuck. The two common types are the four-jaw independent chuck in which each jaw can be adjusted as required to hold out-of-round work or the three jaw self-centring type in which all jaws move in unison. This type is quicker to operate, but the workpiece must be circular. Some of these chucks are power operated by a cylinder which may be either pneumatic or hydraulic. The cylinder is connected to the chuck by a draw bar or a draw tube which allows for the passage of bar stock. There are two basic types of power chuck, i.e. the lever type shown in Figure 3.5(a) and the wedge type. In each type the gripping jaws are held by screws to a sliding part so that the jaws can be adjusted separately to suit varying diameters, the actual jaw movement by the levers being only small. To guard against air failure a valve may be incorporated in the cylinder air line to hold the pressure until the machining is completed. Alternatively, an electrical pressure switch may be installed to operate an alarm bell or warning light if the pressure drops below a pre-set limit.

COLLET CHUCKS

Collect chucks are used for bar work and again can be power operated or tightened to grip the bar stock by a hand lever. Collets are required to be accurately made, for most types rely on their spring-gripping power, and separate collets are required for each diameter of bar which may be round, square or hexagonal depending on the work to be produced. A hydraulic chuck is shown in Figure 3.5(b) (Crawford Collets Ltd), the collet being

Figure 3.5.
(a) A lever-type chuck
(b) A hydraulic collet chuck:
A sleeve B ring C plunger
(By courtesy of Crawford Collets Ltd)

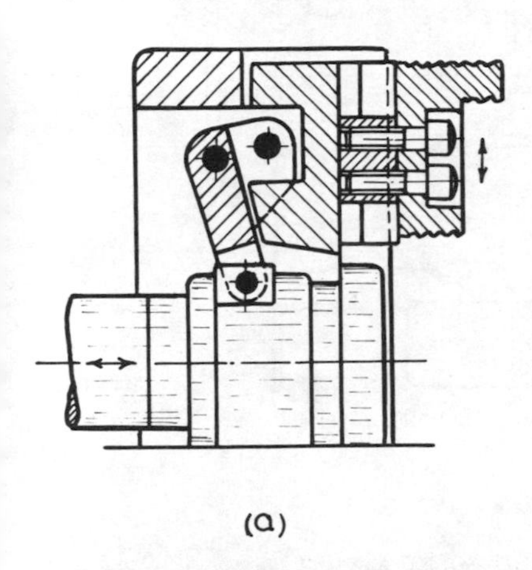

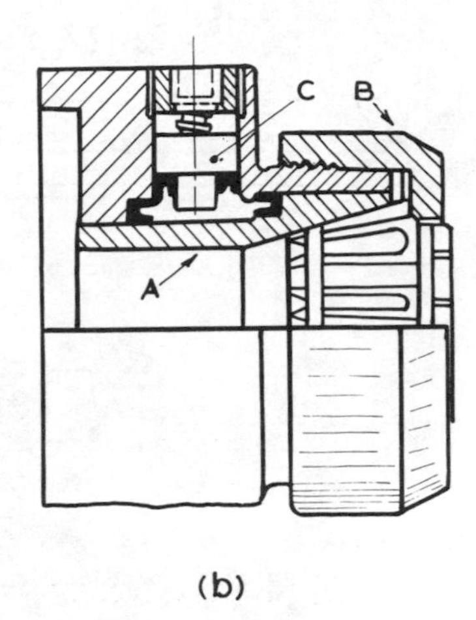

housed in a taper bore at the end of the sleeve A. It is located endwise by the threaded nose ring B, while an annular space between the sleeve and body is filled with oil. It communicates with a cross hole which houses the plunger C; this is threaded to hold a screw with a square-section socket to receive a chuck key. Thus, by the key action, the plunger causes the oil to be displaced and to move the sleeve A to the right, closing the collet by the action of the taper surfaces. When the plunger is moved outwards, springs (not shown) cause the sleeve to move to the left and so to open the collet. The chuck only requires a low torque on the key to exert a very high gripping power on the work.

LEVER- AND POWER-OPERATED COLLET CHUCKS

Figure 3.6 shows a lever-operated design which comprises a sliding cone A at the end of the spindle and actuating toggle levers B which act upon the end of a tube inside the main spindle. The slight endwise movement given to the tube is sufficient to make the spring chuck grip the bar. In some cases a weight and wire rope connection is used to bring the bar forwards, but in the design shown a positive feed is obtained by the lever C which performs the double action of feeding and chucking.

The lever is pivoted with a link to the headstock and slides the sleeve D forked to operate the cone A; this in turn operates the toggles, tube and chuck against the cap E. On releasing the toggles the elasticity of the chuck reacts and returns the tube. When the cone A is moved to the right, the screw in sleeve D draws the ratchet bar F in the same direction but not immediately because of the slot in the bar, thus allowing the chuck to disengage before the bar feed operates. A pawl G connects the bar carrier H

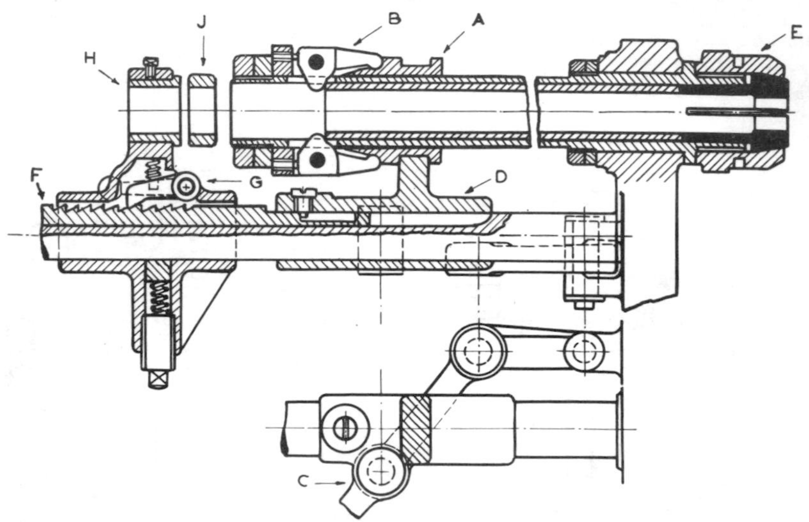

Figure 3.6.
A diagram showing the mechanism to actuate a collet chuck:
A sliding cone *B actuating toggle levers* *C lever* *D sleeve* *E cap* *F ratchet bar* *G pawl* *H bar carrier* *J collar*

with the ratchet bar, the pawl being lifted out of mesh when required to be moved to a new section of bar. A spring-loaded pad under the bar F imparts an even pressure when the carrier is locked. To operate, a collar J is screwed to the bar, and this collar is pushed along by the front of the bush in the bar carrier. A stand carries the end of the shaft in which F is fitted and also supports the ends of the long stock bars being turned.

POWER OPERATION

On automatic lathes the bar feed motion must fit in with the cycle of the machining operations so that, as soon as workpiece has been cut off from the bar after completion, the bar is fed forwards automatically against a dead stop. The bar feed mechanism is shown in Figure 3.7 and comprises two collets. The cycle of operation is as follows.

(1) With the bar held in a friction grip of the feed collet F, the pusher tube P moves the bar forwards until it contacts a dead stop in the turret. This gives the required work length.

(2) At this point the chuck collet C is contacted by the closing tube T as it moves forwards, causing the collet to close and grip the bar ready for machining.

(3) The pusher tube and feed collet move backwards to the original position ready to feed the bar forwards again as soon as the chuck collet is released.

(4) The chuck collet opens for the next forward movement of the bar.

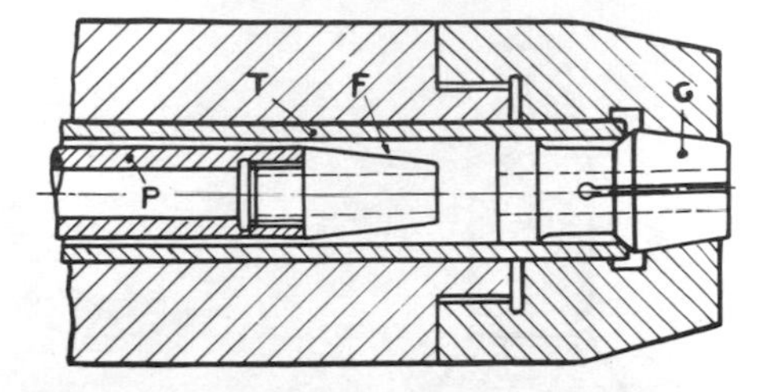

Figure 3.7.
Power-operated collet chucks for an automatic lathe:
C *chuck collet*
F *feed collet*
P *pusher tube*
T *closing tube*

Collect chucks grip the work more truly than other forms of chucks, but the variation in size which can be accommodated is more limited, for a separate collet is required for every work diameter. Also only bright bar can be used, and this must be accurate in size and section, otherwise the friction grip will not function properly. Bar feed and collet chucks are also available for black bar or non-precision bar. However, they differ in that they are hydraulic chucks used both to grip the bar during machining and to grip it whilst it is being advanced after the parting-off operation.

CENTRE WORK

Turning with the work mounted between centres makes it necessary to drill centre holes in each end of the work, and the operation should be performed with a special drill giving a countersink angle of 60°. If the drilling is done in a lathe, the centre drill may be held in a chuck mounted in the tailstock spindle as shown in Figure 3.8(a). Centre drills are delicate tools and are easily broken when feeding the drill if the speed is too low.

Figure 3.8.
Drilling a centre hole by using a chuck in the tailstock

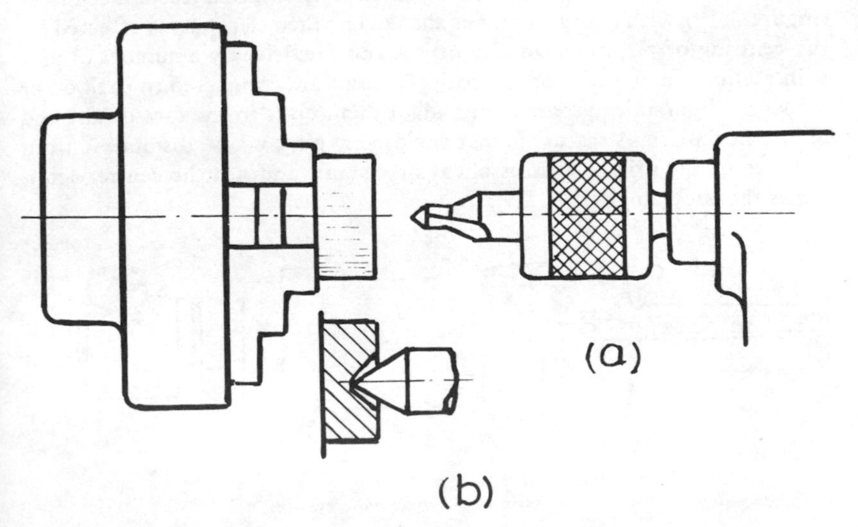

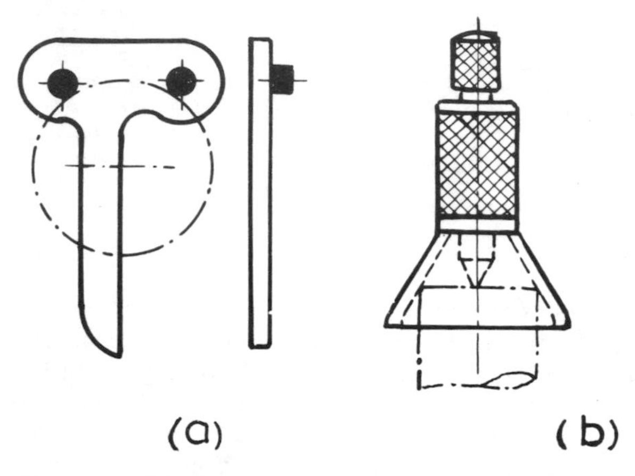

Figure 3.8(b) shows the importance of a correct centre hole and the fault that may occur if a correct drill is not used. Figure 3.9(a) shows a method to centre the end of a bar preparatory to drilling. By using the square, a line is scribed across the end, and then another line at right angles is scribed, the intersection giving the centre. If a cup or bell centre punch is available (Figure 3.9(b)), the operation is simplified.

HYDRAULIC DRIVING CENTRES

When turning between centres the usual method of driving the work is by a pin on the revolving catch plate which contacts a driver on the work. The disadvantage is that, if the workpiece is being machined over its full length, it is necessary to remove the driver and to turn the work around between the centres. A better and time-saving method is to use a Kosta face driver (Figure 3.10) with a locating taper shank. The face clamping is effected by the centring of a spring-loaded centre A and the drive by a number of pins B in a circle on the face of the tool. The pins are connected to each other through a hydraulic system C and adapt themselves to the face on the end of the workpiece. A feature is that the driving pins, whose distance *X* from the face of the driving head is always constant, and not the centre, determines the stop length.

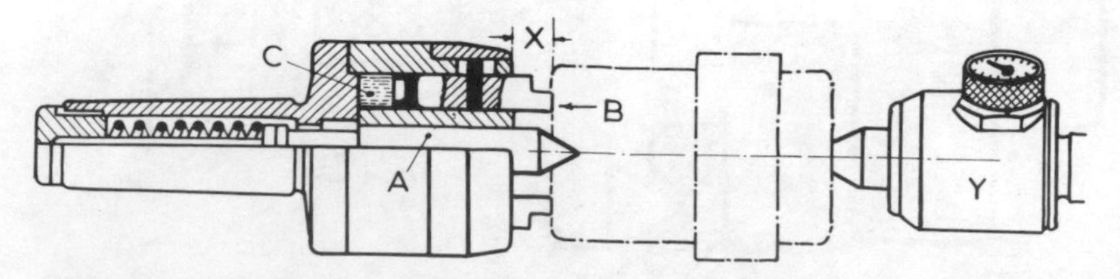

Figure 3.10.
The operation of Kosta hydraulic driving centres:
A spring-loaded centre B pins C hydraulic system

The face driving unit works because the edges of the pins are pressed with sufficient force into the workpiece to transmit a turning moment offering sufficient security, but it is important that the force used should not be excessive. The solution relies on matching the driving unit with a revolving centre which not only shows the holding pressure on the hydraulic gauge but maintains it at a constant level during machining. Thus there are three points to be considered

(1) The amount of axial pressure
(2) The expansion of the work due to heat
(3) The deeper penetration of the carrying pins caused by the cutting process.

MANDRELS

For workpieces that cannot be held in a chuck and have a central hole, the most convenient means of turning is by the use of a mandrel. In its simplest form a mandrel is a hardened and ground shaft with accurately centred ends and is slightly tapered, usually about 0.006 in/ft (0.15 mm/300 mm). A flat is provided at one end to carry a driving dog as shown in Figure 3.11(a). The smaller end is inserted into the bore of the workpiece and is pressed in until it binds tightly. It is advisable to lubricate the mandrel so that seizure does not take place between the mating surfaces. To prevent damage to the centre hole during insertion the ends of the mandrel should be recessed.

The limitation is that a separate mandrel must be provided for every size of hole; so, to reduce this number, various designs of expansion

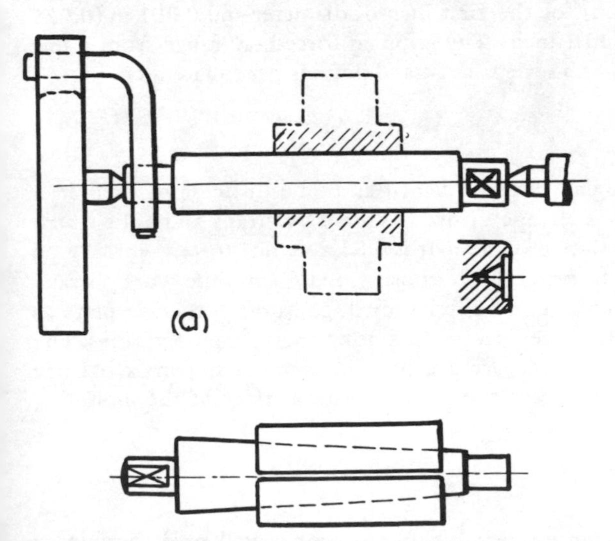

Figure 3.11.
The use of mandrels for holding a workpiece on a centre lathe

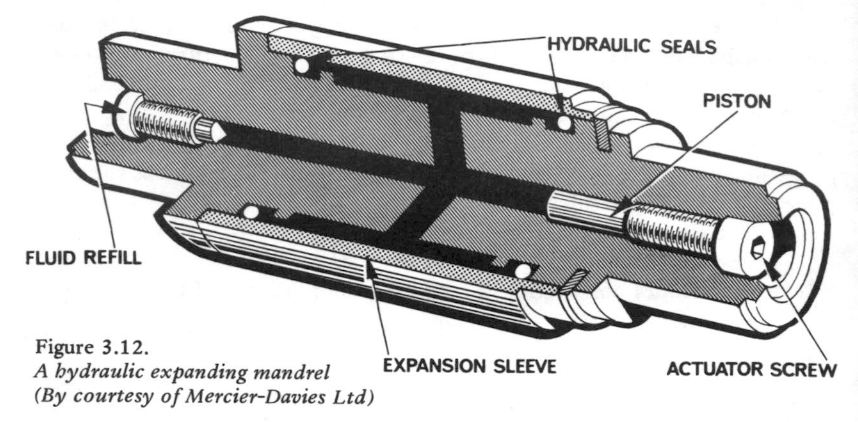

Figure 3.12.
A hydraulic expanding mandrel
(By courtesy of Mercier-Davies Ltd)

mandrels are available. Figure 3.11(b) shows a type with an inner tapering shaft on which is a split bush that can be expanded within limits, by driving in the tapering member. The advantage is that a smaller stock of mandrels is required, for different-sized bushes can be used. In general, however, the solid mandrel is the more accurate of the two types.

A third type (Figure 3.12) is the hydraulic expanding mandrel (Mercier-Davies Ltd) which functions on the principle of expanding metals within their elastic limits under controlled oil pressure. The self-contained hydraulic system holds or positions components on their theoretic centre line, irrespective of 'in-tolerance' errors of ovality, taper or lobing. The mandrel will function within tolerances between 0.00003 in (0.0007 mm) to 0.0005 in (0.0127 mm) according to size. The general rule for expansion is 0.003 in (0.075 mm) for the first inch of diameter and 0.001 in (0.025 mm) for each additional inch. The gripping force may range from a few pounds per square inch to several thousand pounds per square inch.

FACEPLATE WORK

Some castings or forgings are so shaped as to be difficult to hold in a chuck, but they can be clamped on a faceplate by straps and bolts which utilize the holes and slots provided. It is advisable not to rely entirely on these but for safety to provide additional stops against the work to take the pressure of the cut and to nullify centrifugal force. An angle plate as shown in Figure 3.13 is often useful for boring and facing operations. The operation shown is that of boring a bracket, and, as shown, a balance weight is fitted to counteract the out-of-balance effect of the mounting units and work.

USE OF STEADIES

Long slender work if unsupported between centres will tend to whip or bend under the pressure of the cutting action. To prevent this happening,

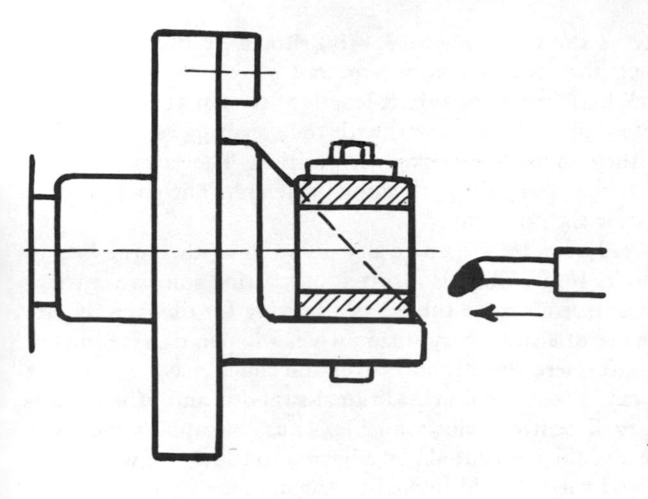

Figure 3.13.
The use of a balanced angle plate for boring operations

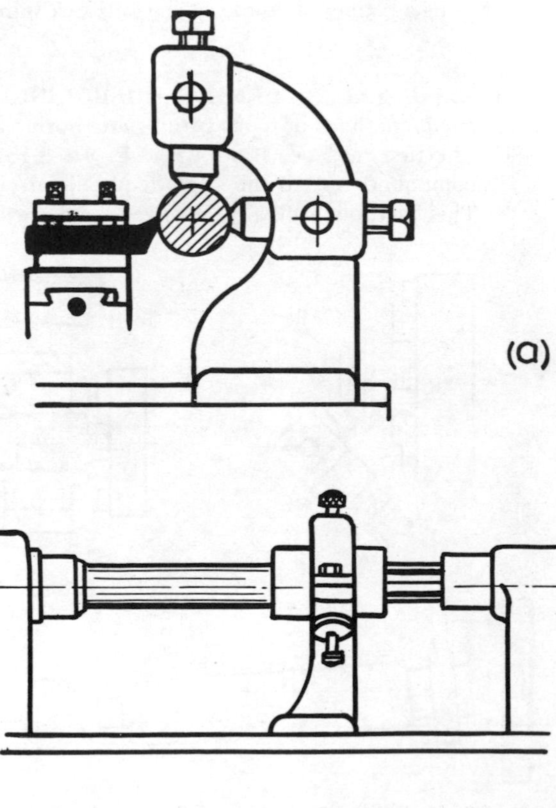

Figure 3.14.
Travelling and stationary steadies for supporting workpieces

a travelling steady as shown in Figure 3.14(a) should be fitted. If bright bar is being turned, the tool can be set to trail behind the steady, but, when turning black bar, first turn a short length of the bar at the tailstock end to the diameter required, and then, with the steady jaws adjusted to touch the work, they should be locked in position. The jaws will then support the work at the point of cut all along the length. The jaws must be kept lubricated during the operation.

A stationary steady can be set up at any point along the lathe bed to support a long shaft. If the shaft is of black bar, a ring somewhat wider than the jaws of the rest must be turned as a bearing for the jaws. Figure 3.14(b) shows the use of a stationary steady when an operation is required on the end of a shaft. Here the distance from the chuck may be too great for machining to take place without additional support, and, if drilling is required, the tailstock centre is not available. Thus, to support the work by means of a steady, the jaws should be adjusted to touch the work until it is running true and then should be locked. Again, plenty of oil must be used between the steady jaws and the revolving work. The operation is that of using a reamer mounted in the tailstock spindle.

METHODS OF TAPER TURNING AND BORING

Three general methods of taper turning and boring are applicable.

(1) The first method, as shown in Figure 3.15(a), is by swivelling the compound rest to the angle required for either turning or boring. This method generally involves hand traverse of the tool and is

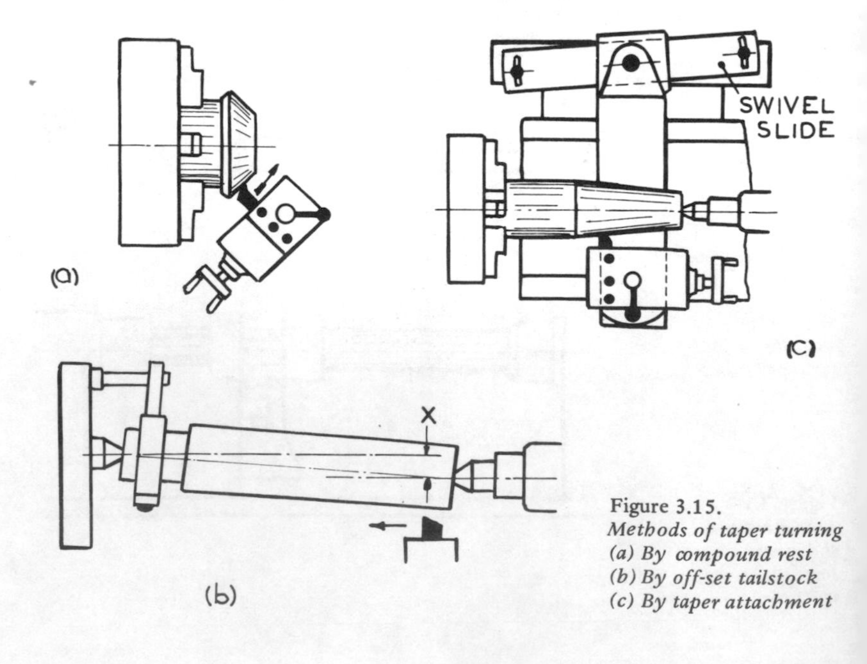

Figure 3.15.
Methods of taper turning
(a) By compound rest
(b) By off-set tailstock
(c) By taper attachment

limited to the length of movement of the top slide, but it has the advantage that taper surfaces of any angle can be machined.

(2) The method shown in Figure 3.15(b) is by off-setting the tailstock centre. The drawback is that now the centre points are not in alignment, so that the centres are subjected to uneven wear and strain. Thus the method is limited to slow tapers on long work.

To find the amount of the off-set X, if the taper is T in/ft and the length of work is I in then

$$X = \frac{T \times I}{24} \text{ in}$$

If the included taper angle is θ,

$$X = I \times \tan\left(\frac{\theta}{2}\right) \text{ in}$$

(3) If the lathe is fitted with a taper turning attachment (Figure 3.15(c)), then more accurate tapers, either external or internal, can be produced than by the two preceding methods. By the use of this attachment, the lathe centres are not taken out of alignment. For all taper turning operations, it is essential that the cutting edge of the tool be mounted exactly on the centre line of the work.

This can be seen from Figure 3.16. To turn the taper shown, the tool T would be traversed back a distance X while moving along the work. Assume, however, that the tool was mounted at the height A, then the tool would again move back the same distance X, but the large end would be undersized as shown by the chain curves, if we assume that the small end is the same in each case. For example, it will be apparent that a lathe centre could not be turned to a sharp point if the tool was not set on the centre height.

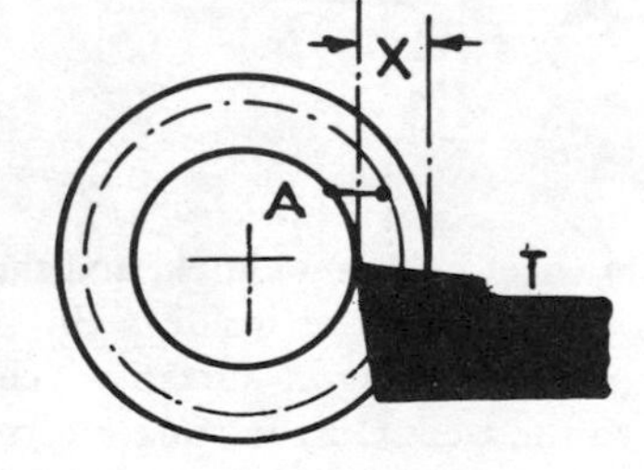

Figure 3.16.
A diagram showing why the tool point must be on centre for taper turning

FORMING OPERATIONS

Taper diameters can be produced on a workpiece by forming as at Figure 3.17(a). There must be sufficient rigidity on the tool rest to prevent vibrations developing, for the taper is formed by a straight-in cut with the full width of the tool. Forming operations are not restricted to taper shapes

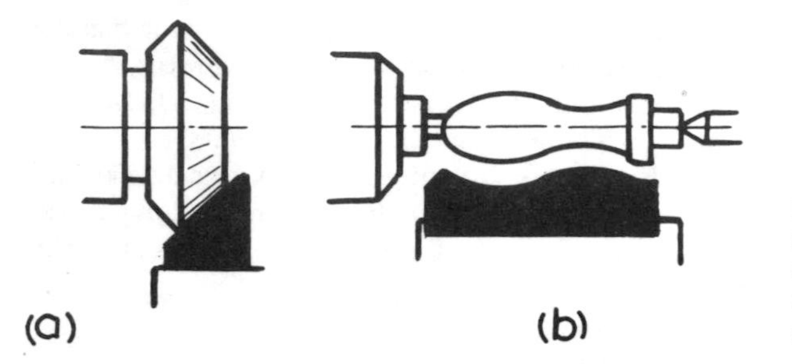

Figure 3.17.
Forming angular curved surfaces by the straight-in method

but may produce curves, as for example a shaped handle as shown in Figure 3.17(b). A very efficient method of producing tapers or other contours is by tracer-controlled copying; this is described in Chapter 7.

PROFILING

Straight-in forming is restricted to surfaces of comparatively short and rigid formation, and for long profiles a single-point tool is used. Not all tapers to be produced lie in the same plane of direction as the lathe centres, for some surfaces taper in a direction across the bed. When profiling in this direction the taper bar must be supported on the bed, by the tailstock or, as in the case under consideration, by a taper bar on the turret (Figure 3.18).

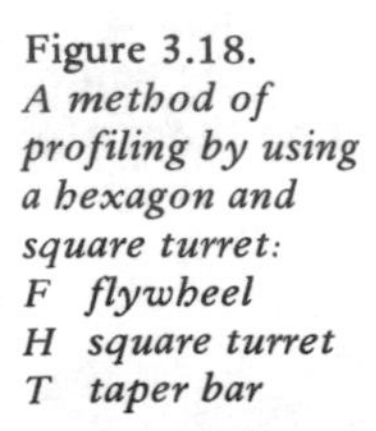

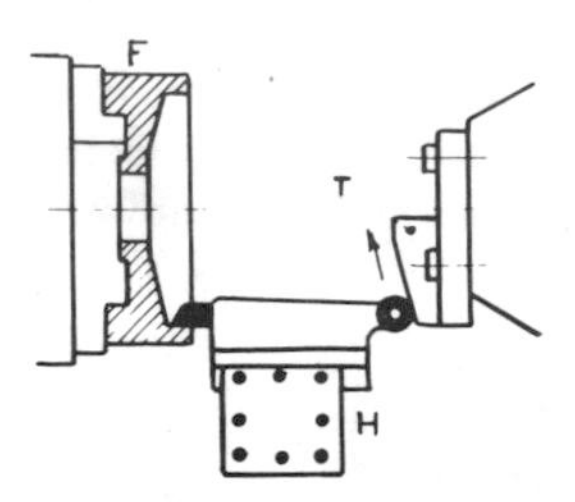

Figure 3.18.
A method of profiling by using a hexagon and square turret:
F flywheel
H square turret
T taper bar

The component F is a flywheel held in a chuck, and the required process is that of producing a taper face on the inside web. The taper bar T is shown bolted to a face on the hexagon turret which is kept rigid by clamping the turret slide to the bed. On the square turret H is clamped a slide carrying the tool holder which carries a cutting tool at one end and a roller at the other, so that, when the surfacing feed is engaged, the point of the tool follows a path dictated by the angle of the taper bar, thus reproducing the desired shape on the flywheel web. This method of profiling gives a good surface finish and is particularly applicable on heavy work. For example, large pistons may have the closed end radiused by the same set-up or by an alternative arrangement of mounting the taper bar on the square turret and the roller slide on the hexagon turret.

SCREW-CUTTING METHODS

Most modern lathes are fitted with a feed gear box which has a wide range of gear changes so that the majority of screw thread pitches can be obtained, but in other cases the gear box may be restricted to a few changes, so that change wheels must be used for screw-cutting. The full set comprises 22 gears commencing with 20 teeth and rising by steps of five teeth to 120, including two wheels of 40 teeth for cutting screws of the same pitch as the lead screw of the lathe.

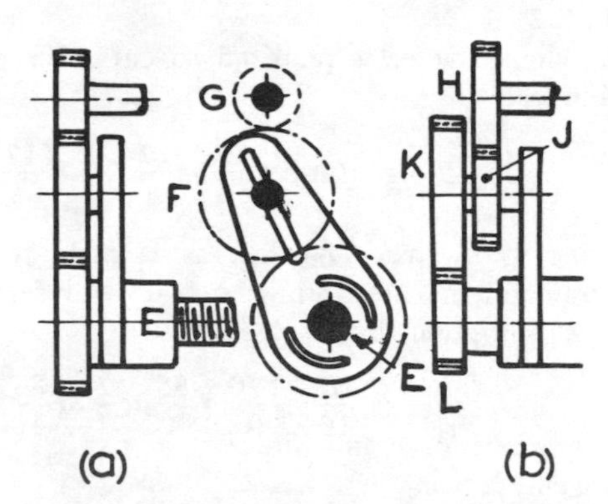

Figure 3.19. *Methods of mounting change wheels for a screw-cutting operation*

Two methods of arranging the gears are used: Figure 3.19(a) shows a single train, and Figure 3.19(b) a compound train. The latter is used when a large speed reduction ratio between the headstock spindle and the lead screw must be obtained.

Two terms are used in describing a screw thread, pitch and lead. Pitch denotes the distance from a point on one thread to a corresponding point on the next thread, whereas lead is the distance that a nut travels per revolution of a screw. With ordinary screws such as bolts and nuts, the thread is single.

SELECTION OF CHANGE WHEELS

The ratio of the change wheels may be found from the lead of screw to be cut divided by the lead of lathe screw.

EXAMPLE 1

Find the change wheels to cut a screw ¾ in lead on a lathe having a lead screw of ½ in lead.

$$\text{Ratio} = \frac{3/4}{1/2} = \frac{3}{4} \times \frac{2}{1} = \frac{3}{2}$$

multiplying this by some number gives suitable change wheels such as

$$\frac{3}{2} \times \frac{10}{10} = \frac{30}{20}$$

the 30-tooth wheel is on the headstock stud G and the 20-tooth wheel on the end of the lead screw E. These two would be connected by an intermediate wheel F, say 40 teeth, thus

$$\frac{G}{F} \times \frac{F}{E} \text{ or } \frac{30}{40} \times \frac{40}{20}$$

the intermediate wheel has no effect upon the speed ratio.

EXAMPLE 2

Find the change wheels required to cut a screw $2\frac{1}{8}$ in lead. Lathe lead screw ½ in lead.

$$\text{Ratio} = \frac{2\frac{1}{8}}{\frac{1}{2}} = \frac{17}{8} \times \frac{2}{1} = \frac{17}{4}$$

Multiplying by 5 gives $^{85}/_{20}$, but as a single train this would impose a heavy strain on the 20-tooth wheel; it is preferable to use a compound train, thus multiplying by 10 gives

$$\frac{17}{4} \times \frac{10}{10} = \frac{170}{40} = \frac{85}{40} \times \frac{2}{1} \text{ or } \frac{85}{40} \times \frac{60}{30}$$

the arrangement being

$$\frac{H}{J} \times \frac{K}{L} \text{ or } \frac{\text{driver}}{\text{driven}} \times \frac{\text{driver}}{\text{driven}}$$

To check the train,

$$\frac{85}{40} \times \frac{60}{30} \times (\tfrac{1}{2} \text{ in lead screw}) = 2\tfrac{1}{8}$$

MULTIPLE-THREADED SCREWS

Where the rapid movement of a nut along a screw is required and where the use of a large pitch would weaken the screw, multiple threads are used. The pitch can then be fine, but the lead can be increased to give the required movement. As an example, the information given on a drawing might read ¼ in pitch, 1 in lead and four threads (or starts). Thus the cutting tool would be ground to suit the pitch, but the change wheels mounted to give 1 in traverse for every revolution of the spindle.

The method of cutting a double-threaded screw is shown in Figure 3.20(a), which indicates that one screw is cut, and is then followed by the second thread which occurs centrally between the parts of the first thread as in Figure 3.20(b) (this is usually known as a two-start thread).

There are several ways of dealing with the spacing problem. One method is by means of a micrometer dial on the compound rest, so that, after one thread has been cut, the tool can be reset in the next position as shown by the broken lines.

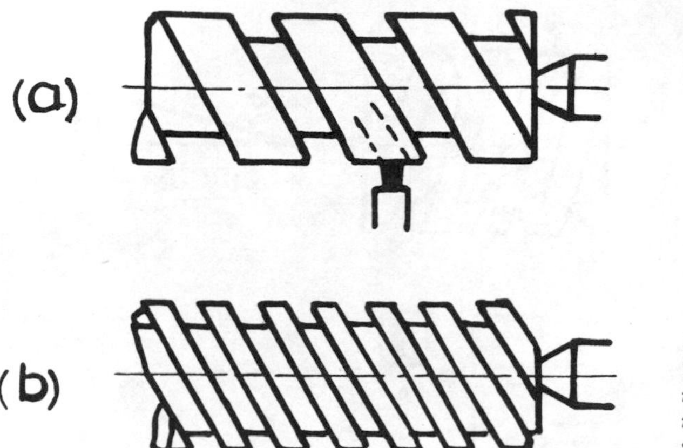

Figure 3.20.
The procedure in machining a multiple-thread screw

THE SCREW-CUTTING DIAL

This dial is found on the lathe apron, and when cutting screws the lead screw nut can be engaged at any point and another cut can be taken so long as the number of threads to be cut is divisible by the number of threads per inch of the lead screw. With odd numbers, however, cross threading will result, so that a screw-cutting dial should be used. This consists of a dial A, (Figure 3.21) connected to a worm wheel B in mesh with the lead screw, so that, if the saddle is stationary, the revolving lead screw acting as a worm causes the dial to rotate.

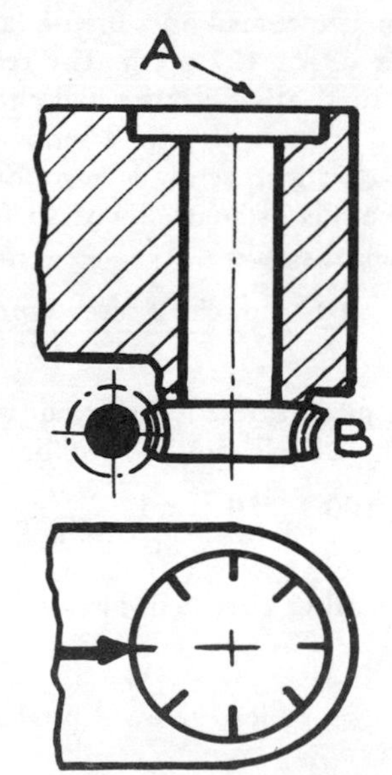

Figure 3.21.
A screw-cutting dial used to prevent cross threading:
A dial
B worm wheel

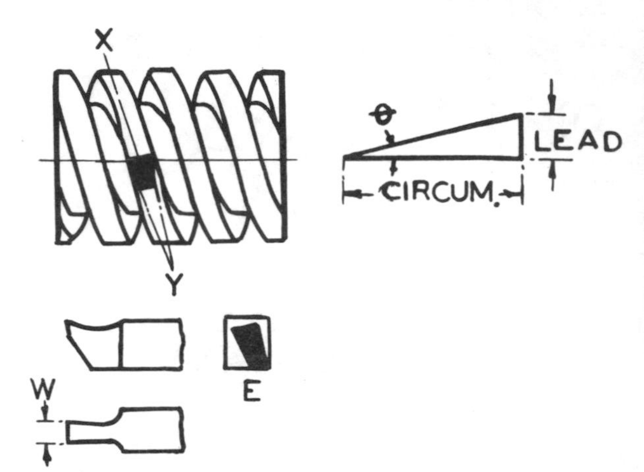

Figure 3.22.
The shape and setting of a tool to cut a square thread

When the nut is engaged and the tool commences its travel, the dial remains stationary with one of the graduations opposite the arrow. After the cut has been completed and the saddle has been returned to its starting point with the nut disengaged, the dial commences to revolve again, and when one of the graduated lines comes opposite the arrow, the nut can be re-engaged with the assurance that the tool will follow the same cut. The number of teeth in the worm wheel divided by the pitch of the screw equals the number of graduations on the dial.

METRIC THREADS

The problem of cutting a metric thread on a British lathe is solved by using a translating gear of either 63 or 127 teeth. The reason is that there are almost 25.4 mm in 1 in, so that, if a lathe had change wheels mounted $^{100}/_{254}$ and a lead screw had a pitch of 1 in, a screw of 1 mm pitch would be cut. As the usual pitch of a lead screw is ½ in, or less, then 127 teeth can be used, although 63 teeth is often used with sufficient accuracy.

To find the change wheels the rule is

$$\frac{\text{driver}}{\text{driven}} = \frac{10}{127} \times \frac{\text{pitch of screw (mm)}}{1}$$

For example, if a 10 mm pitch screw is to be cut on a lathe with a lead screw of ½ in pitch, then the wheels required will be

$$\frac{10}{127} \times \frac{10}{1} = \frac{100}{127} \text{ or } \frac{50}{127} \times \frac{63}{30} \text{ compound train}$$

For lathes with lead screws other than ½ in pitch the following rule can be applied:

$$\frac{\text{driver}}{\text{driven}} = \frac{5}{127} \times \text{threads per inch of lead screw} \times \text{pitch (mm) of required screw}$$

The rule to find the change gears for cutting British threads on metric lead screws is

$$\frac{127}{\text{thread to be cut} \times \text{metric screw (mm)} \times 5}$$

EXAMPLE OF MACHINING BORING MACHINE SPINDLE

FIRST SETTING

First operation

Cut off to length + ¼ in; face ends and centre both ends.

Second operation

Grip end E in chuck and support end A by tailstock centre. Turn diameters A to B to finished grinding size. Turn diameters C and D parallel to largest diameter.

Third operation

Undercut where indicated at each change of diameter.

Fourth operation

Using compound rest, turn taper D to grinding size; hand feed.

Fifth operation

Using taper turning attachment machine taper C to grinding size.

SECOND SETTING

Reverse end position of spindle gripping A in chuck; turn E.

THIRD SETTING

Support spindle end in stationary steady.

First operation

Drill taper hole to small diameter by using drill in tailstock.

Second operation

Bore Morse taper to reamer size, by using either swivelled compound rest or taper-boring attachment. Cut recess at bottom of taper.

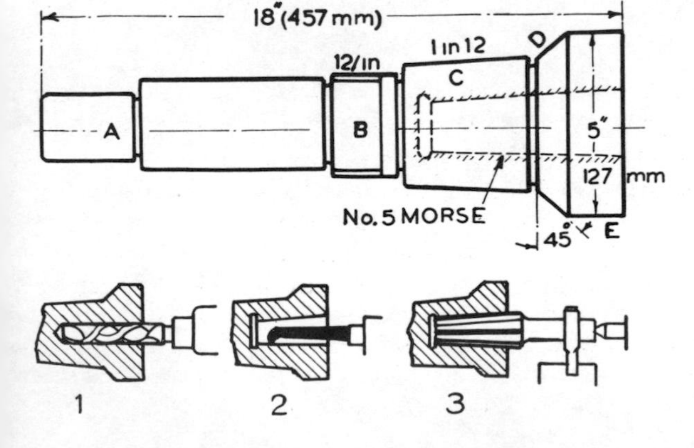

Figure 3.23. *Machining operations for turning and boring a machine spindle*

Third operation
With Morse taper reamer supported on tailstock centre and with carrier on shank contacting the compound rest, feed forward, making sure that the lathe centre does not leave centre hole in reamer.
Fouth operation
With a centre plug in the Morse taper bore, mount the spindle in the lathe, and strike the thread on part B.

Figure 3.23, parts 1, 2, 3 show the tools used for machining the Morse taper bore.

4

DRILLING AND BORING OPERATIONS AND MACHINES

SENSITIVE DRILLING MACHINE

Small high-speed machines of simple construction are known as sensitive drilling machines, for the feed to the drill is by hand only. Heavier-type machines are usually of vertical construction with elevating and swivelling tables, or of the radial type in which the drill can be swung and located over the workpiece. The construction of the spindle mechanism of a sensitive drilling machine is shown in Figure 4.1; the hand pressure for drilling is applied by lever A to rotate a pinion which gears into the rack on the spindle sleeve. The driving pulley is mounted on a ball-bearing sleeve so that no belt pull comes on the spindle. The only connection between the pulley and the spindle are two keys to cause spindle rotation as it slides through the sleeve. The end pressure of the drill is taken by double-purpose ball bearings located at the bottom of the spindle; these also form ball journal bearings for the spindle rotation.

The driving and feed mechanism for heavier machines is shown in Figure 4.2. The driving shaft carries a friction clutch A, and from either side gearing connects to the first shaft, two connecting wheels at one side and three at the other giving a choice of forward or reverse motion. Nine speeds are then obtained by sliding gears on two shafts meshing with gears on an intermediate shaft. From this gear box the drive terminates by driving the spindle through two spiral gears, so that with the driving shaft revolving at 400 rev/min the fastest spindle speed is

$$\frac{400}{1} \times \frac{45}{50} \times \frac{34}{19} \times \frac{35}{33} \times \frac{14}{24} = 400 \text{ rev/min approximately}$$

The feed motion is derived from the spindle pinion with 26 teeth, connecting to the 86-tooth wheel on the first shaft. Four speeds are obtainable: these are selected by sliding the member F and are engaged by the clutch C. The transmission continues to the worm and wheel motion and terminates with the rack pinion on the worm wheel shaft meshing with the spindle socket. The fastest feed is thus

$$\frac{26}{86} \times \frac{27}{27} \times \frac{1}{100} \times \frac{2}{1} \times \frac{22}{7} = 0.018 \text{ in/rev } (0.457 \text{ mm/rev}) \text{ of spindle}$$

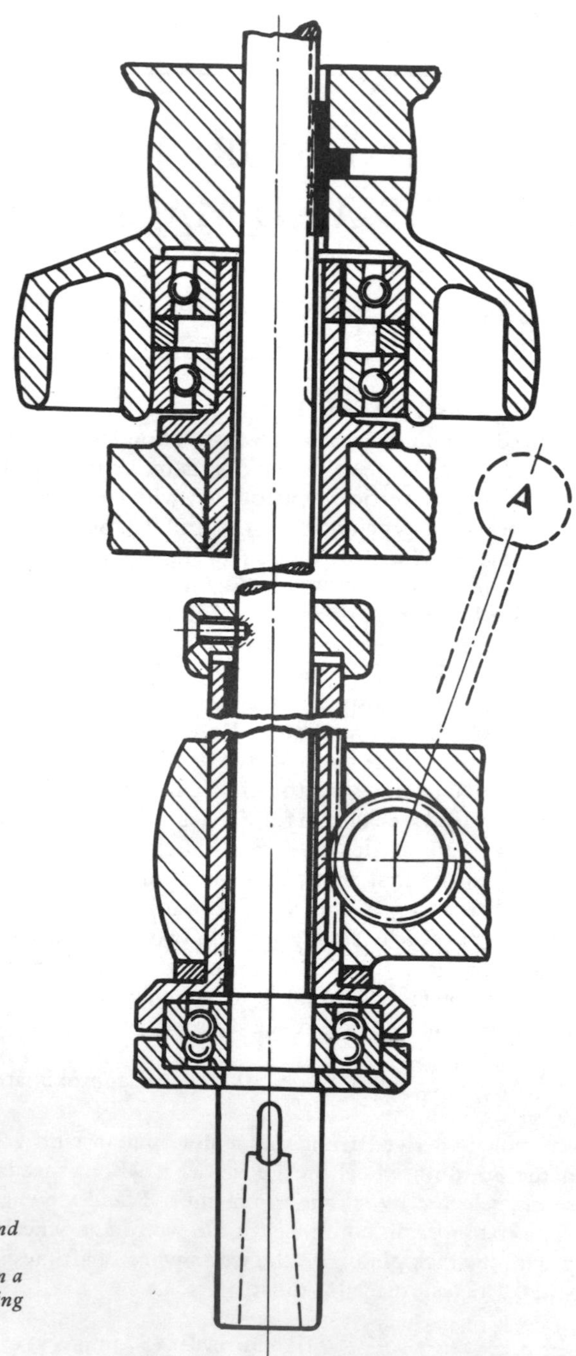

Figure 4.1. *The driving and hand feed mechanism on a sensitive drilling machine:*
A lever

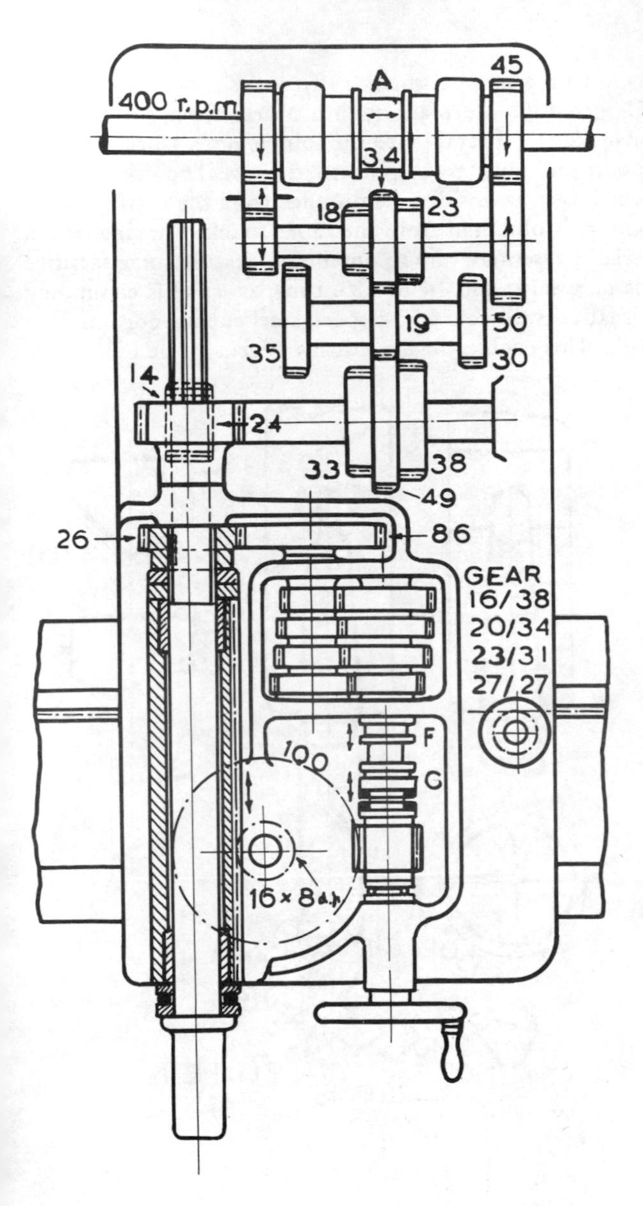

Figure 4.2.
Change speed and feed details for drilling machine saddle:
A friction clutch
C clutch
F feed change

The spindle rotates inside a steel sleeve with rack teeth cut on the outside for the spindle traverse. The sleeve has bronze bushes at the top and bottom and ball thrust washers to take the end thrust of the drill. Adjustment is by a pair of locknuts at the top of the sleeve.

TWIST DRILLS

Small drills are made with a parallel shank to fit in a chuck, but as shown in Figure 4.3(a); larger drills are made with a Morse taper shank to fit either in sleeves or sockets or directly into the spindle nose. During cutting a drill gets no support and must rely upon its stiffness. The driving is by means of a tang which is at a considerable distance from the cutting edges, although some support is obtained from the taper shank. For this reason toughness of material is essential, and to obtain this feature some sacrifice of cutting speed is necessary, and 14 to 18% tungsten steel is commonly employed. An alternative is to weld a high-speed steel cutting portion onto an alloy steel shank. This enables higher cutting speeds to be used while

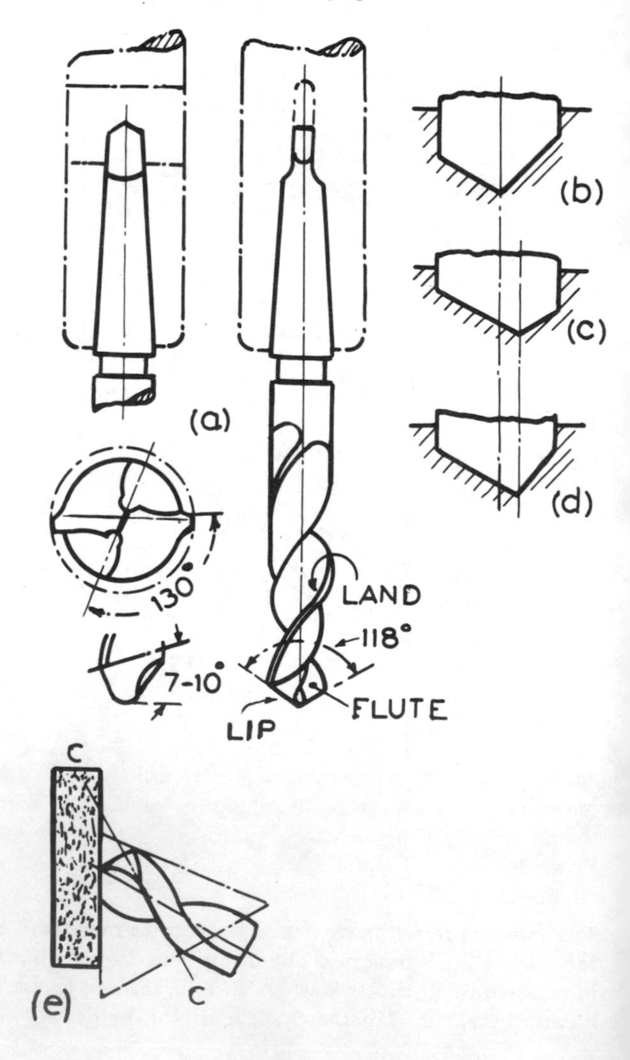

Figure 4.3.
Details of twist drill giving angles and grinding method

retaining a strong driving and locating portion of the drill. For special drills cemented carbide tips are available. The cutting edges terminate on two 'lands' which give body clearance. Another clearance is on the diameter of the drill itself which tapers about 0.0075 in (0.19 mm) on the diameter as it is ground away. This allows the drill to run in a deep hole without binding. The helix of the flutes give the rake angle, and the two flutes together leave a central web, forming the chisel edge on the end of the drill. This edge must be forced through the metal so that point thinning is employed to reduce the length of the edge.

The lip angle is made 118° for general purposes, for a smaller angle would weaken the point and would increase the length of the cutting edges. A greater angle makes it difficult for the drill to centre itself and to commence cutting. A clearance angle of 7 to 10° is provided.

INCORRECT GRINDING

Drills are usually made with two grooves to form a double-threaded spiral, so that cuttings are forced up the grooves and clear of the hole. Unlike most other cutting tools, drills do not derive guidance from the machine spindle, although accurate location is essential, but from their own cutting edges. If the radial components at the two cutting edges are equal, the drill should follow a straight path, but often, through lack of uniformity in the metal, a drill will take the line of least resistance and will follow a curved path.

These conditions can be worsened by incorect grinding. One defect is shown in Figure 4.3(b) where the cutting edges are ground to different angles. Here nearly all the work is thrown on one edge, forcing the drill to one side so that it cuts too large a hole. If the cutting edges are ground to equal angles (Figure 4.3(c)), but of different lengths, the point of the drill is off-centre, so that the drill is revolving on one axis and the point on another. Again the hole will be oversize, and drill breakage may take place. The worst condition (Figure 4.3(d)) is when the cutting edges are of both unequal angles and unequal lengths, with the point off-centre. With a large drill damage may not be restricted to the drill but may extend to the machine spindle.

When a drill is being ground it is held at an angle to the face of the grinding wheel and is rotated so that the face of the wheel comes into contact with the entire surface at the back of each cutting edge. In grinding by hand, this movement may be irregular, but with machine grinding the movement is mechanically controlled. The motion required is shown in Figure 4.3(e), and the clearance should gradually increase from the outer circumference towards the point of the drill. The rotation of the drill is about an axis CC which is inclined from the face of the grinding wheel somewhat less than the axis of the drill. When a drill is ground in this way, the end is given a conical surface, the apex of the cone being above the point of the drill, as indicated by the broken lines.

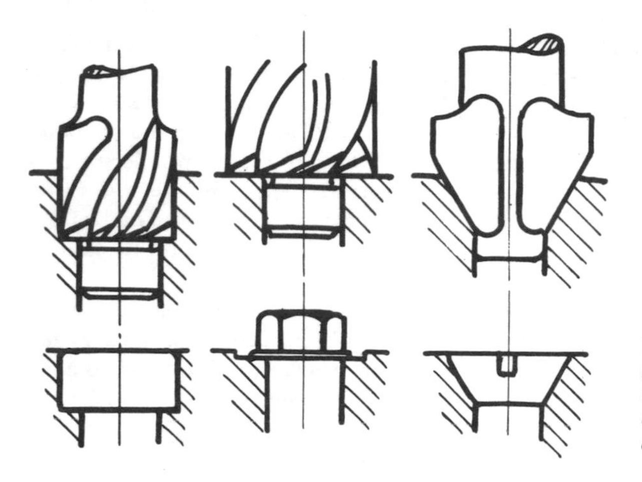

Figure 4.4. *Supplementary tools used in conjunction with drilling operations*

SPECIAL AND SUPPLEMENTARY TOOLS

Core drills are available with three or four flutes; their purpose is to enlarge cored holes in castings. The drills have flat ends and cannot be used for drilling solid metal. As no cutting is done by the point, the flutes are made rather shallow with thickening of the web and stiffening of the drill. Counterboring and countersinking tools are employed to enlarge a hole or to provide a machined face for a set screw or collar. Three examples are shown in Figure 4.4. Clearance is provided on the end of the cutting teeth only and not on the cylindrical portion, as it does not cut. To facilitate sharpening, a groove is provided between the guide and the body. The amount of clearance is 4 to 5°.

DEEP HOLE DRILLING

There are three main problems encountered in drilling holes

(1) The tendency of the drill to run out of true.
(2) The difficulty of removing cuttings.
(3) The heating of the tool and work.

In ordinary drilling with the tool revolving and the work stationary, unless a drill is started exactly true on the required centre, there is an ever-increasing tendency for it to deviate still further from its true course with a possibility of drill breakage. This is shown in Figure 4.5(a) which shows that the wedge action at B is continually forcing the drill to the left as the work moves along the line BC. On the other hand, as indicated in Figure 4.5(b), if the work revolves against a stationary drill, and again the drill is running out of true, the point will trace a circle of diameter XX. This tends to bend the drill equally all around the circumference of X, so that the drill resistance pulls towards the centre of the work where the pull on it is the least. Thus, for deep hole drilling, the axiom is for the work to

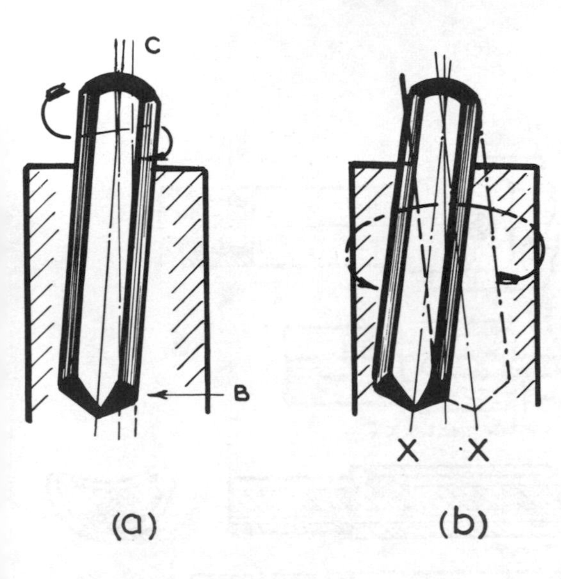

Figure 4.5. *Diagrams illustrating the problems of deep hole drilling*

revolve against a stationary drill when feasible.

The feed motion is generally applied to the drill, and for small-diameter drills trouble can occur unless the drill is withdrawn at intervals from the bore to clear cuttings. High-pressure coolant is required to clear the chips. For drilling from the solid, on such examples as machine tool hollow spindles, a type of spade drill can be used, (Figure 4.6(a)). Holes up to 6 in (152 mm) in diameter can be drilled from the solid. The cutter is held in the end of a short head, locating itself by two lips and being locked by a screw. The cutting edges are notched to break up the chips and are bevelled at 17°. The bar is screwed into the main boring bar which is provided with a central hole for the coolant which enters the head through a similar central hole and then deviates into two channels to supply both sides of the cutter.

For very accurate holes, for example rifle drilling, a 'D' drill is used. The main feature of the drill is that, given a true start, accuracy of size and alignment will be maintained, so that it is customary to drill a hole for a short distance, to bore to size with a single point tool and then to commence drilling with the 'D' drill.

Figure 4.6(b) shows the drill which derives its name from its shape, which allows cuttings to escape easily but leaves sufficient tool section to ensure that it must follow a true path. The drill is fixed to a tubular boring bar through which coolant is pumped at a high pressure to pass through the drill and to remove cuttings. The tubular bar should be of the same section but slightly less than the cutting portion, which should taper slightly from the cutting end to prevent any tendency to seize up in the hole. 'D' bits 1½ in (38 mm) in diameter usually feed about 0.001 in/rev (0.025 mm/rev).

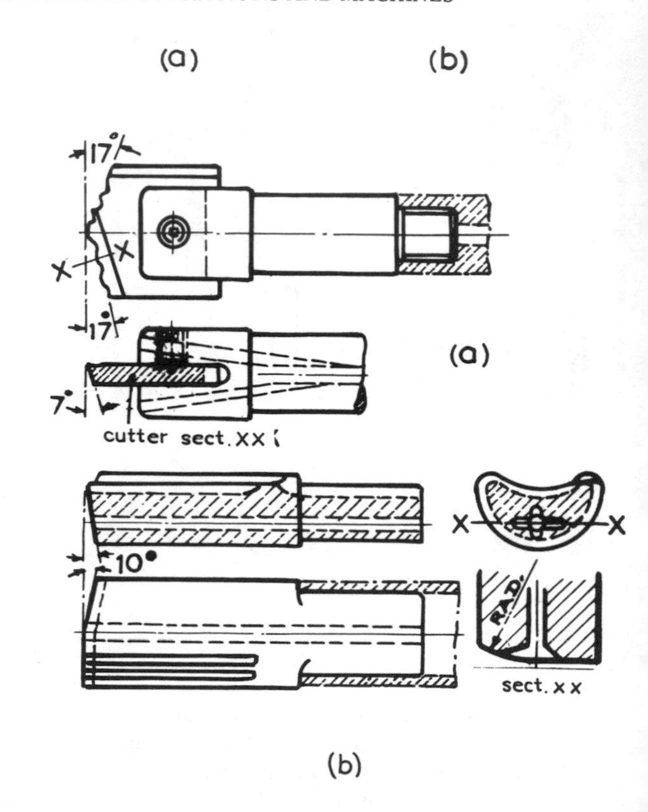

Figure 4.6.
Deep hole drilling tools
(a) Spade drill
(b) 'D' drill

DRILLING WITH CARBIDE

Figure 4.7 depicts the range of ejector drilling tools (Sandvik U.K. Ltd) which has enabled the advantages of cemented carbide to be used for drilling holes from 18 to 65 mm in diameter. An example of one of these tools indicating how cuttings are washed away is shown in Figure 9.6. At the shank end there is a multi-start square thread to enable the cutter head to be attached to the outer of a pair of co-axial tubes, the coolant being delivered under pressure to the annular space between the tubes. It passes over guide pads for lubrication and to the carbide tips to remove the cuttings.

Holes with diameter to depth ratios of 1 to 100 can be drilled, and in general the bore diameter can be held to within 0.1 mm and the straightness to within 0.05 mm in 300 mm.

BORING TOOLS AND OPERATIONS

The horizontal boring machine is one of the most versatile of machine tools, for the traverse motions to the spindle, to the saddle on the column and to the table in the longitudinal and transverse directions enable

Figure 4.7.
Ejector drilling tools with carbide edges
(By courtesy of Sandvik U.K. Ltd, Halesowen)

operations such as milling to be performed. This is in addition to boring, facing and drilling. The saddle construction of a typical machine is shown in Figure 4.8. In order that wear does not occur on the boring spindle, it slides in a revolving socket and receives its feed motion from the rack which is clamped at any suitable position along the bar and is traversed by rack pinion and worm gearing. A set of boring bars to carry cutters and to fit the Morse taper hole in the spindle is provided, the opposite end of the

Figure 4.8.
The construction of a boring machine saddle showing transmission elements

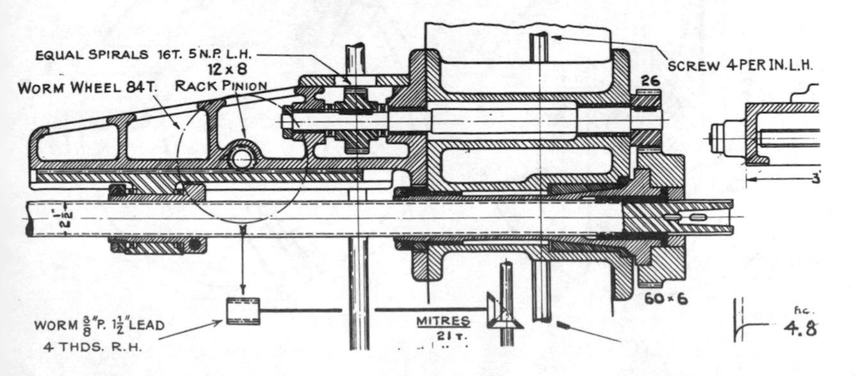

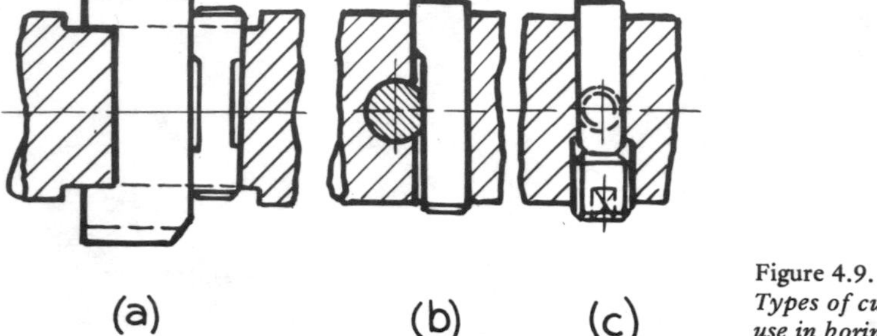

Figure 4.9.
Types of cutters for use in boring bars

bars being supported by bushes in the outer support. The driven wheel on the front of the spindle socket is provided with tee slots to which boring or milling heads can be bolted just as on a faceplate.

BORING CUTTERS

Suitable slots are spaced out on the boring bars for the cutters, and for roughing operations double cutters can be used as shown in Figure 4.9(a). The bar has two flats milled on it, while the cutter is made with lips to fit over the flats after passing through the bar. Different sizes of cutters will thus centre themselves every time and will ensure that both edges are cutting alike. A wedge is used for locking the cutter, preferably at the front so that the pressure of the cut is taken against the solid bar.

For truing a bore, a single-point tool of round or rectangular section is essential. Round cutters are generally used, as they are easy to assemble and adjust. Two methods of holding are shown in Figure 4.9(b) and (c). The first cutter is provided with a flat on the side against which a flattened taper pin is driven to hold the cutter in position and to prevent rotation.

Figure 4.10.
Boring heads for machining large-diameter holes

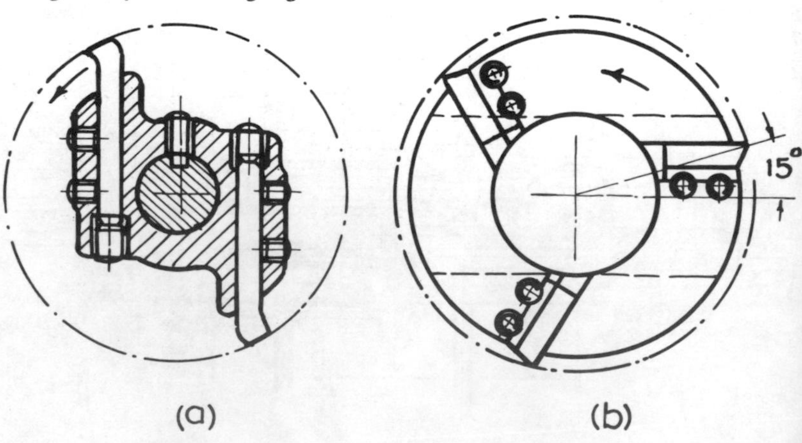

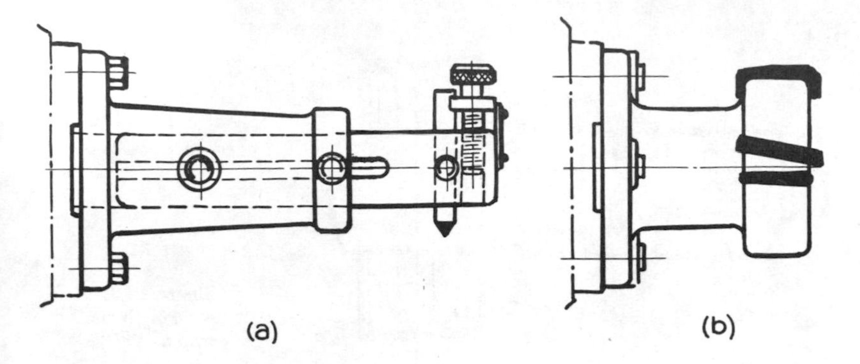

Figure 4.11.
An extension tool holder and milling head driven from faceplate

The second method shows screw adjustment, while locking is by a small screw acting on the side of the cutter. These cutters are often set at an angle, but, with multiple tools machining up to shoulders, they are better set perpendicular to the axis of the bar, for angular cutters are more difficult to grind and set.

For the boring of large-diameter holes, boring heads can be keyed or bolted to the boring bar as in Figure 4.10(a) and (b). The first carries two tools of circular section with separate adjustment for each, while

Figure 4.12.
A facing head attached to the machine faceplate

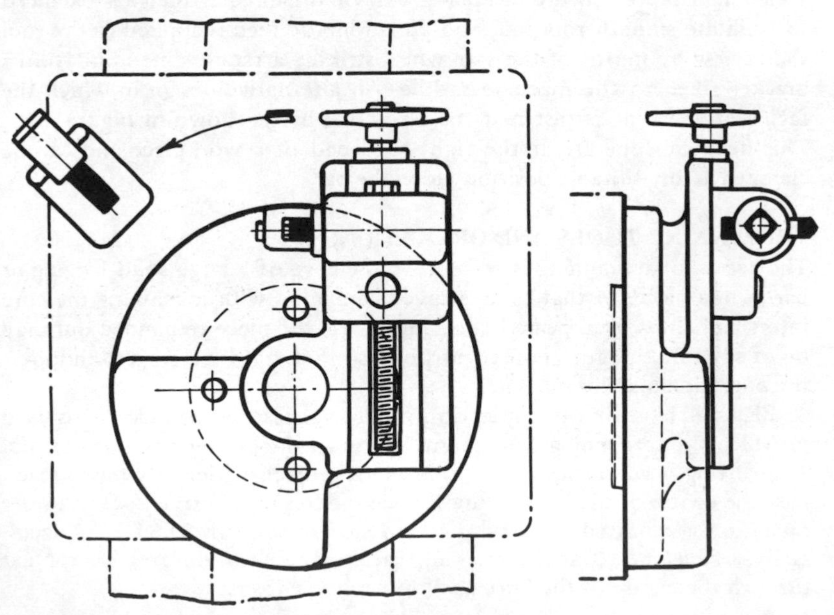

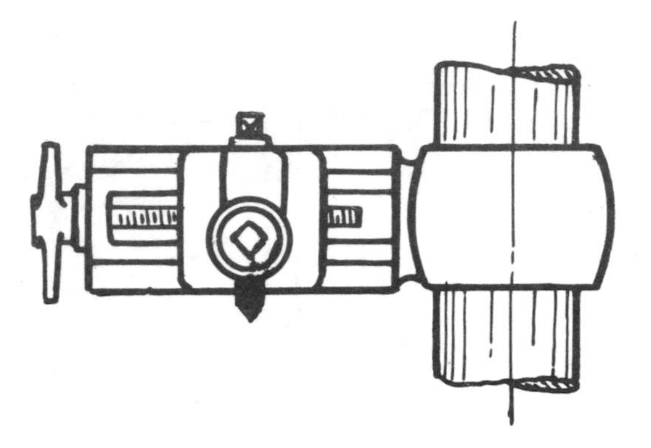

Figure 4.13.
A facing head attached to the boring bar

Figure 4.9(b) shows negative rake cutters of cemented carbide. The negative angle gives firm support and reduces the shocks if intermittent cutting takes place. Up to 25° negative rake is not excessive if accompanied by side and end clearance.

For boring and facing operations which employ the faceplate of the driving wheel instead of a boring bar for holding the tools, extension tool holders can be used as shown in Figure 4.11. Screw-cutting can be carried out on the boring machine if the table longitudinal screw is coupled to the feed shaft and if change wheels are supplied. Figure 4.11(a) shows a tool for this purpose, and Figure 4.11(b) a similar mounting for a milling cutter. Many machines have an automatic facing head, but Figure 4.12 shows an attachment bolted to the faceplate. Part of the circular flange is balanced to facilitate smooth rotation, and an automatic feed is applied to the tool slide screw by means of the star wheel striking a tappet extending from a bracket fixed to the machine saddle. An alternative design in which the facing attachment is fastened to the boring bar is shown in Figure 4.13. This head can operate at the right-hand end of a workpiece and can be clamped at any suitable position along the bar.

TREPANNING TOOLS AND OPERATIONS

The action of trepanning is to remove the core of a large solid forging or bar in one piece, so that time is saved compared with machining the core into small chips by repeated cuts. Moreover, the piece trepanned out may be of suitable size for another purpose, so that from a salvage standpoint the operation may be valuable.

Figure 4.14 shows the operation when a vertical boring machine is used in which the trepanning equipment locates on the boring machine spindle. The cutting head carries three tools which project sufficiently far to reach past the centre of the bore so that, when the forging is turned over and the operation is repeated, a useful piece of steel is cut out. The feed is necessarily a fine one to avoid breaking the cutters, and lubricant is supplied through the centre of the boring spindle to the cutters.

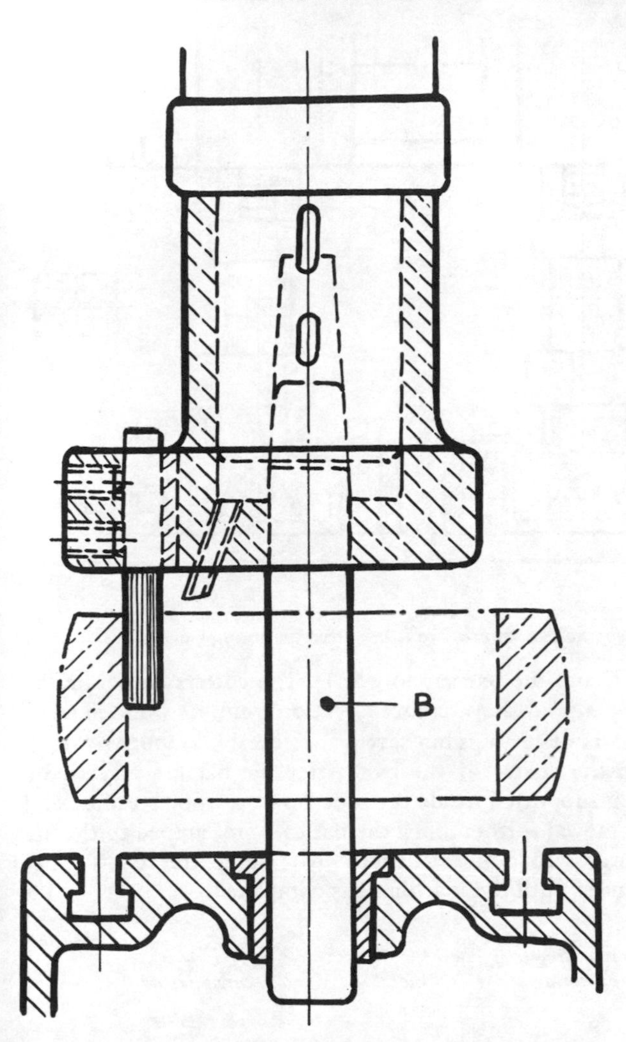

Figure 4.14.
A trepanning operation on a vertical boring machine:
B boring bar

To finish the bore to size, a boring bar B is fitted to the machine spindle and is piloted in a bush in the table. The finishing head is keyed to the bar under the trepanning head which is not removed while boring takes place. The actual trepanning heads are shown in operation in Figure 4.15 depicting the machining of diesel engine connecting rods using a duplex boring machine so that two rods are machined simultaneously. Suitable fixtures are used to hold the rods, and the metal saved by the trepanning operation is shown on the right- and left-hand sides of the fixtures. In this instance the piece saved is 10 in (254 mm) diam. and 6 in (152 mm) thick.

For trepanning long bars the method can be understood from Figure 4.16, which shows the workpiece at A, the hollow trepanning bar at B,

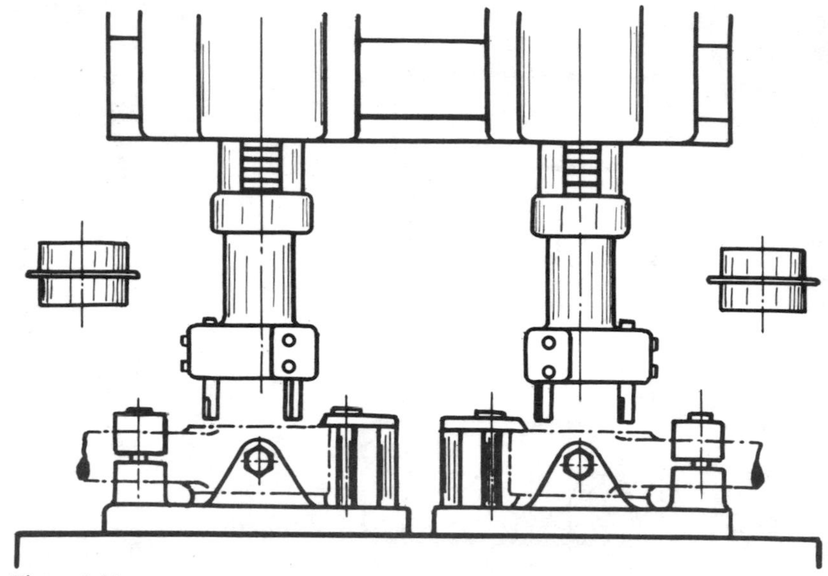

Figure 4.15.
The trepanning diesel engine connnecting rods on a duplex boring machine

the metal saved at C and the bar support at D. The cutters are set on the end of the tubular bar, and they remove a section from the work equal to their width in the form of cuttings but save a considerable amount of stock which passes down the centre of the bar. After the bar has traversed a certain distance, it is supported inside the bore on hard wood steadies, and the driving head D moved further along the bar to be reclamped to the bar and to be fed forwards for another length. After each traverse, further supports are introduced until the machining is completed.

Figure 4.16.
An arrangement for trepanning long bars:
A workpiece B trepanning bar C metal saved D driving head

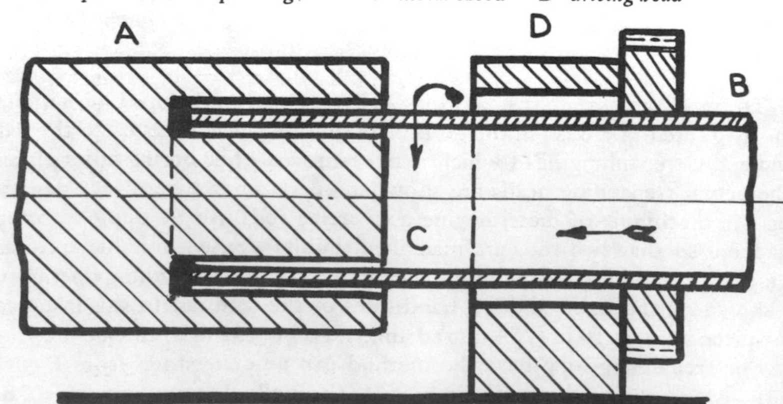

FINISHING BORES BY REAMING

To obtain the requisite accuracy in the sizing of holes, a reaming operation is usually performed; while with careful alignment of the machine and reamer, accurate results may be obtained, it is preferable that some floating action between work and tool should be present. This feature of a tool that is able to centre itself accurately in relation to a previously machined hole eliminates any misalignment present in the machine and ensures better results than can be obtained with a rigid tool, so that, if ordinary reamers are used, a floating action is usually obtained by a flexible mounting of the reamer holder.

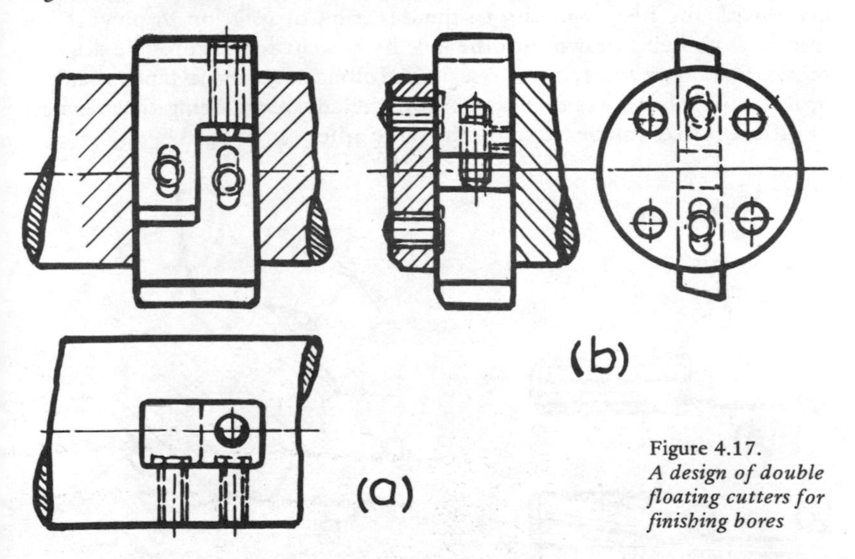

Figure 4.17.
A design of double floating cutters for finishing bores

Good results can be obtained by double floating cutters, the only limitation being that sizing is a matter of trial and that adjustment is over a small range. The type shown in Figure 4.17(a) is used for supported boring bars, the two cutters being held from falling out of the slot by screws inserted in the side of the bar. These screws allow sufficient movement of the cutters to obtain a floating action, while size adjustment is by the screw passing through one cutter. The type shown in Figure 4.17(b) can be used for shank bars and is easy to manufacture by having a separate end cap.

A range of adjustable floating reamers of the type described are manufactured by David Brown Ltd, Huddersfield, but have the refinement that size can be set to a dial reading and fine adjustment can be made by nut and screw. The range of adjustment is also wide so that the number of reamers to cover a range of sizes is less than when rigid reamers are employed. The recommended angle of top rake for these floating cutters is 3° for cast iron, 15° for mild steel, 10° for hard steel, 20° for white metal, aluminium and Duralumin, and 0° for bronze, brass and gun metal.

The hole should be bored leaving a reaming allowance of 0.006 in

(152 mm) on the diameter. The recommended speeds for cutting are for cast iron 20 ft/min (6 m/min), for mild steel 35 ft/min (10.6 m/min), for medium-carbon steel 20 ft/min (9 m/min), and for white metal, aluminium, phosphor-bronze and Duralumin 35 ft/min (10.6 m/min).

Recommended feeds range from 0.19 in (0.9 mm) to 0.36 in (4.7 mm) according to the diameter of the hole.

STANDARD REAMERS

These may have straight flutes as shown in Figure 4.18(a) or helical flutes (Figure 4.8(b)) to give a better shearing cut. Helical fluted reamers are made with the helix opposite to the direction of rotation to prevent the reamer from being drawn into the hole by the cutting action. The diagram shows hand reamers, but for use in a drilling machine a taper shank is provided. Shell reamers fit on an arbor, the advantage being that reamers of different sizes can be fitted on the same arbor as required.

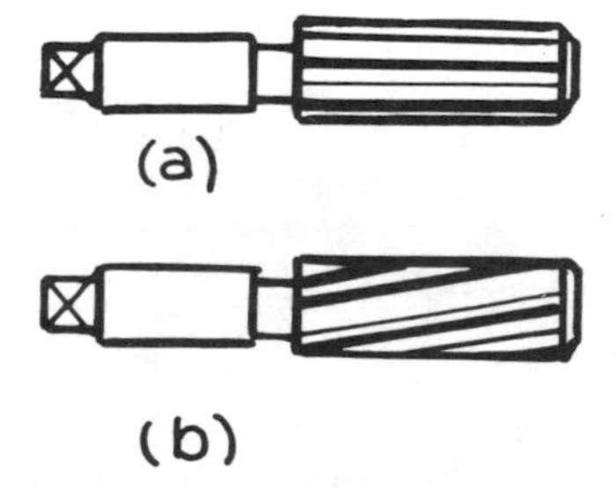

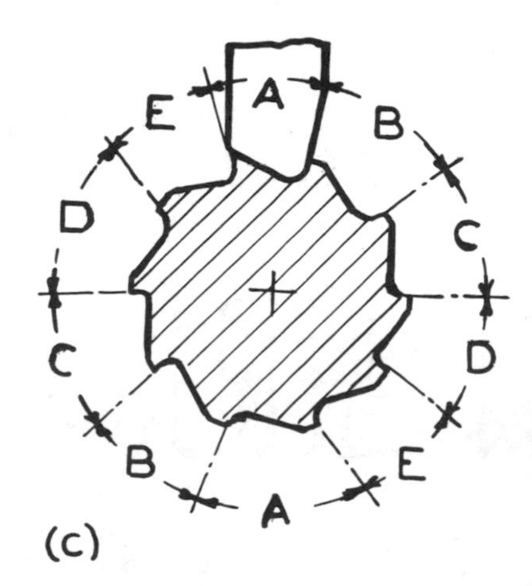

Figure 4.18. *Reamers and angular cutters for machining teeth*

The cutting angle of a tooth is often made with a zero or negative rake angle up to 5°. There are, however, certain reasons why a positive angle may prove of advantage. The motion of a reamer is that of a forward movement plus a rotary motion, the two giving a helical motion to the cutting edges. There is another motion on a tangent line; this is a line which is formed on the diameter of the reamer at the point of contact of the tooth, thus giving a force acting along the tangent line which is at right angles with the straight line that passes through the tangent point of contact and the centre of the circle of operation.

If the tooth was formed parallel to the straight line, the angle would be neither positive nor negative. If the tooth face formed an angle less than 90° with the tangent line, on the side from which the force acts, the negative rake would not give the power to resist the force acting along the tangent line; also the force along the tangent line tends to act outwards

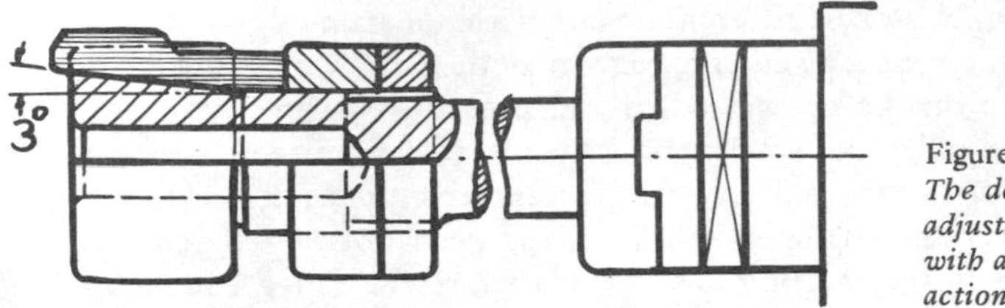

Figure 4.19. *The design of an adjustable reamer with a floating action*

from the centre. If a reamer is made with a positive rake of 3°, the power is diverted in an inward direction, and yet the reamer has no tendency to dig in.

Figure 4.18(c) shows a form of reamer easily cut with a cutter of angular side. Hand reamers should never have less than 6 teeth, the number increasing with the diameter. Odd numbers of teeth give better support, but an even number facilitates measurement. There are two alternatives to irregular spacing of the teeth. In the first, as shown, the flutes are irregular spaced, but opposite cutting edges are diametrically opposite, so that measurement is easy. In the second, the teeth are cut in an irregular spacing so that no two teeth are opposite. The advantage of the latter is that there is no position in which teeth will coincide after a partial revolution, whereas in the first case, if it does start to cut a 'cornered' hole, it will continue to do so.

ADJUSTABLE REAMERS

As shown in Figure 4.19, the adjustment for size is effected by means of the check nut which forces the blades up the inclined guides. As it is difficult to ensure that the face of the adjusting nut will maintain its concentricity owing to wear of the threads, a plain collar is interposed between the nut and the blades. This collar is a push fit on the body of the reamer with both sides true with the bore, so that any movement of the collar ensures that all the blades are moved exactly the same distance irrespective of any inaccuracy of the nut. The slots for the blades are made to taper 3° to the axis with a dovetail section of 12° angle on one side. The hole in the front of the reamer body is for lightening purposes as it is important to avoid excessive weight with a floating reamer. The floating action in this case is by means of the Oldham coupling shown.

SCREWING TAPS

The types used depend upon the thread type required, but a general division is for two classes, hand and machine taps. The important points in manufacture include the effective diameter, the major diameter, the core diameter, the pitch and the angle of thread form. There are B.S.I. standards for all thread forms.

The flutes form both the cutting edges for the threads and the grooves for the cuttings. Hand taps are usually made with three or four flutes, and Figure 4.20 shows different forms with the same nomenclature for the various features. The cutting edges may be radial, but a circular groove provides rake on the cutting edges and prevents clogging by cuttings on reversal. The relief on hand taps is just sufficient to give free cutting. Figure 4.20(a) and (b) indicates the features mentioned; however, for fine threading operations (Figure 4.20(c)) taps need not be relieved, but the width of the cutting portion is small relative to the flutes, and so friction is small and clogging is prevented.

Taps with an odd number of flutes are difficult to measure, but there is an advantage in a three-flute tap in that it gives a good cutting action and ample chip clearance. Even single-flute taps are used on aluminium, while a two-flute tap (Figure 4.20(d)) will give good results on all soft metals. A backed-off tap with straight flutes is guided by as many line contacts as the tap has flutes, whereas a spiral-fluted tap is supported around a greater proportion of its periphery and obtains better guidance. The best form of tap relief is shown in Figure 4.20(e) where the centres of the radii are on an inclined plane and where each radius is greater than the tap diameter. Taps made with this form of relief cut well without clogging and do not tend to produce oval holes.

As a guide to tap relief for cutting mild steel, relief values are given in Table 4.1, relief for cast iron and bronzes being increased 50%:

Table 4.1.

Diameter (in(mm))	Relief value (in(mm))
0.375 (9.5)	0.002 (0.05)
1 (25.4)	0.004 (0.10)
2 (50.8)	0.007 (0.178)
3 (76.2)	0.009 (0.23)

DIAMOND BORING TOOLS

These tools possess the same advantageous features for producing accurate holes as those described for turning operations in Chapter 1. Similarly, the conditions for using the tools must be the same, i.e. a machine free from vibration, a belt drive to the spindle from a motor isolated from the machine and a very accurate spindle with either pre-loaded anti-friction bearings or under hydro-dynamic control. These conditions are fulfilled by the fine-boring machine, an example being shown in Figure 4.21. The success of the fine-boring process is based upon the facility with which a single-point tool running at high speed with a fine feed will produce a round hole.

Machines are built with one unit head only, or possibly six if required; these are built virtually around the workpiece. The one shown, however, comprises four heads, the boring operation being from rigid short bars

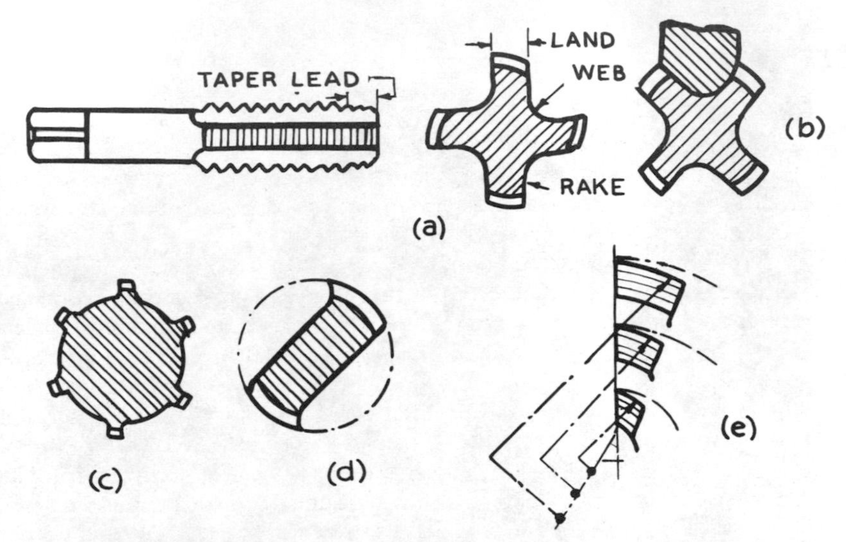

Figure 4.20.
Shapes of flutes on screwing taps for various materials

holding either carbide or diamond tools for machining the holes in a sewing machine body. Owing to the close spacing of the bores, after machining the large holes from both ends, the saddle is indexed to bring the small bores in line with the second spindles. The table traverse is by hydraulic operation giving a smooth movement with the ability to run the table against a dead stop for accurate facing operations.

TOOL SETTING

Compared with diamond turning tools, boring cutters are simple, the round-nose diamond tip being brazed onto a round-section shank which can be clamped in the boring bar. Figure 4.22 shows the tool geometry as recommended by the B.S.I., and the sizes are detailed in Table 4.2.

Table 4.2

Bore diameter (in)	Bore diameter (m)	Diameter D of tool shank (in)	Diameter D of tool shank (m)	Primary clearance angle α	Secondary clearance angle β	Nose angle γ
$\frac{1}{2}$ to $\frac{3}{4}$	13 to 19	$\frac{1}{4}$	6.3	18	23	80
$\frac{3}{4}$ to 1	19 to 25	$\frac{5}{16}$	8.0	15	20	80
1 to $1\frac{1}{2}$	25 to 38	$\frac{5}{16}$	8.0	15	18	80
$1\frac{1}{2}$ to $2\frac{1}{2}$	38 to 63	$\frac{5}{16}$	8.0	10	None	80
$2\frac{1}{2}$ to 3	63 to 76	$\frac{3}{8}$	9.5	7	None	80

Figure 4.21.
An example of a fine boring machine using diamond tools

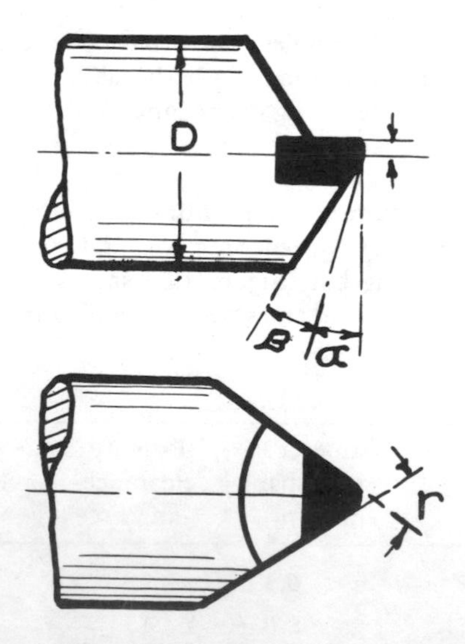

Figure 4.22.
A boring cutter with a round-nose diamond tip

Figure 4.23(a) indicates the requirements for a true bore. The diameter *A* should be the maximum compatible with swarf clearance. A value of 80% of the work diameter is a good average; the effective length *B* should be the minimum to clear the workpiece length *K* and should not in any case

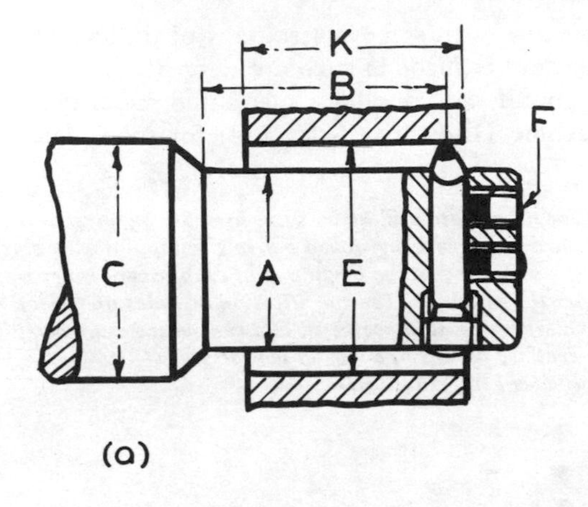

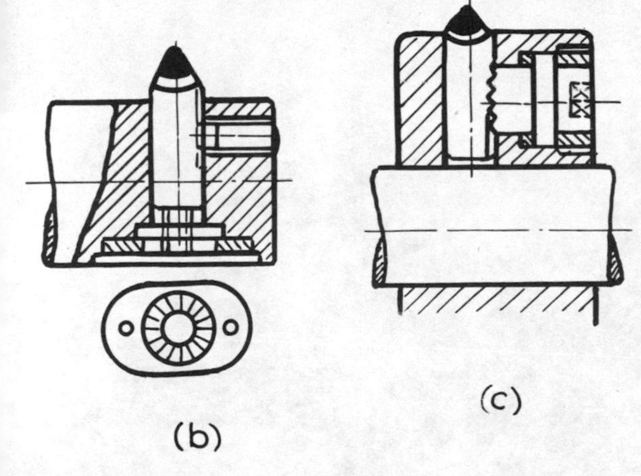

Figure 4.23. *Examples of boring tools showing means of size adjustment*

exceed five times the bar diameter, while the holding diameter *C* should not be less than the working diameter. Two retaining screws with copper pads are preferable, and the tool bar angle is usually 90°, but 45° is employed for machining blind holes. Tool adjustment for diameter is by a fine thread screw. This is shown in Figure 4.23(b) where the screw is of 40 threads/in and a nut with micrometer graduations is held in a fixed axial position but is free to rotate in the bar although retained by a plate. Differential screws which simultaneously engage threads in the boring bar and the shank can be used. If the pitches are 0.025 and 0.02 in, then for one revolution of the screw an adjustment of $0.025 - 0.02 = 0.005$ in is obtained to the diamond.

The scroll-type principle can be used as shown in Figure 4.23(c) where the scroll teeth mesh with rack teeth on the side of the tool bit. One graduation of the dial may equal 0.0005 in.

This chapter terminates with some illustrations of drilling and boring machines. No machine tool is made in such a variety of size and types as the drilling machine, but if we consider a specialized radial type, Figure 4.24 illustrates a machine (Industrial Sales Ltd) for tube plate drilling

Figure 4.24.
A radial drilling machine for tube plate drilling by using an indexing mechanism. It uses three drills simultaneously and with a grooved bar to give rapid positioning of the head. The workpiece is positioned at the beginning of each row of holes by means of the table mounted under the machine. The specification includes pre-selection of speeds and feeds, power clamping of all movements and centralised controls. Pitch-controlled tapping is achieved by means of a master lead screw
(By courtesy of Industrial Sales Ltd, Wilmslow)

which uses three drills simultaneously with a grooved bar under the saddle to give rapid positioning of the drilling head. The workpiece is located at the beginning of each row of holes by means of indexing the table by using the handles and dividing plate below the table. There is pre-selection of speeds and feeds, power clamping of all movements and centralized control. Pitch-controlled tapping is achieved by means of a master lead-screw. There is an outer support to the arm, which carries a bearing for one end of the grooved bar.

Figure 4.25.
Multiple drilling of typewriter frames with vertical and horizontal spindles

Figure 4.25 shows a multi-spindle drilling machine with a vertical head carrying 30 drills for operating on typewriter frames. All holes are drilled simultaneously, the drills being supported in guide bushes located in the jig plate under the spindles. Two additional small drilling units are used to drill holes in the side of the frame, and these units are operated by compressed air from the flexible pipes shown. The feed motion to the vertical head is operated hydraulically.

Diagrams have been shown of facing attachments for the horizontal boring machine, but many machines, as shown in Figure 4.26, incorporate an automatic facing head into the general design. Thus simultaneous boring and facing opeations can be performed on workpieces such as flanged pipes, including turning over the outside diameter of the flange by using a tool in the tool rest of the facing head and a feed motion to the work table. There are independent drives to the boring bar and the facing head, the latter rotating at about one-third of the speed of the boring bar, owing to working on larger diameters.

Figure 4.26.
An automatic facing head of a horizontal boring machine

5

MILLING MACHINES AND OPERATIONS

MILLING PROCESS

Unlike machine tools which use single-point tools to cut metal, the operation of milling is based upon the rotation and relative travel of a multi-tooth cutter, the metal removal being shared by several teeth at each revolution of the cutter as it passes over the work. Special conditions sometimes require a single blade only, and this is known as fly cutting. The operations performed produce results similar to those produced by shaping or planing but by using special cutters or attachments more complicated work can be produced at one pass.

The various types of milling machines have been described in Chapter 2, but, if we consider the machine illustrated in Figure 5.1 (T.S. Harrison & Sons Ltd), eight spindle speeds ranging from 45 to 1,000 rev/min are available, which are selected by the levers on the side of the column, while the machine is started or stopped by the ball-ended lever. Eight table feeds are available by pick-off gears, while the feed motion is started and stopped by the push buttons seen under the right-hand end of the table. The feed can be tripped electrically at any point of the table traverse by setting a dog in the tee slot on the front of the table.

For batch production purposes the machine is available with a fully automatic cycle which comprises a rapid traverse longitudinal motion to the table in addition to the eight feed rates; this rapid traverse is at 280 in/min (7,620 mm/min). The control panel shown in Figure 5.2 now carries four push buttons these comprising start, stop, 'inch' and reverse respectively. The rapid traverse motor is shown on the left of the table, while the feed motor is behind the push-button panel. The automatic cycle is controlled by the three dogs on the side of the table which engage switches on the centre panel. Thus the operator can engage the rapid traverse movement (the work approaching the cutter at high speed), then slow down to the appropriate feed for milling and finally return at high speed to the loading station.

TOOTH CUTTING ACTION

Because of the intermittent action or the limited period of engagement of each tooth with the workpiece, the removal of metal in milling is

Figure 5.1.
A plain horizontal milling machine with push-button control
(By courtesy of T.S. Harrison & Sons Ltd, Heckmondwike)

Figure 5.2.
The Harrison milling machine with an automatic table cycle

accomplished by the separation of small individual chips. The rotary motion of the cutter and the relative travel of the workpiece combine to produce a chip whose cross section varies continuously as the direction of motion of a tooth changes with respect to the direction of motion of the workpiece. Chip thickness is at a maximum at the instant that these motions are most nearly perpendicular to each other, and substantially zero when they are parallel.

The action of a milling cutter can be seen from Figure 5.3(a). The general rule is that the feed motion of the table must be against the cutter, thus producing a comma or trochoid form of chip, for the cut commences at A and terminates at B. Actually there results a slippage area in which the metal is consolidated by the pressure of the cutting edge; this generates a large frictional force which results in wear of the cutting edge. Because of this feature an alternative method in which the feed acts in the same direction as the cutter is employed. This is known as down-cut or climb milling, shown in Figure 5.3(b), and is suitable for cutting hard materials, for no slippage area is formed, but cutters with special angles are required, and there must be no backlash between the feed screw and nut, so that special backlash eliminators must be fitted as otherwise the cutter may drag the table forwards and cause cutter breakage.

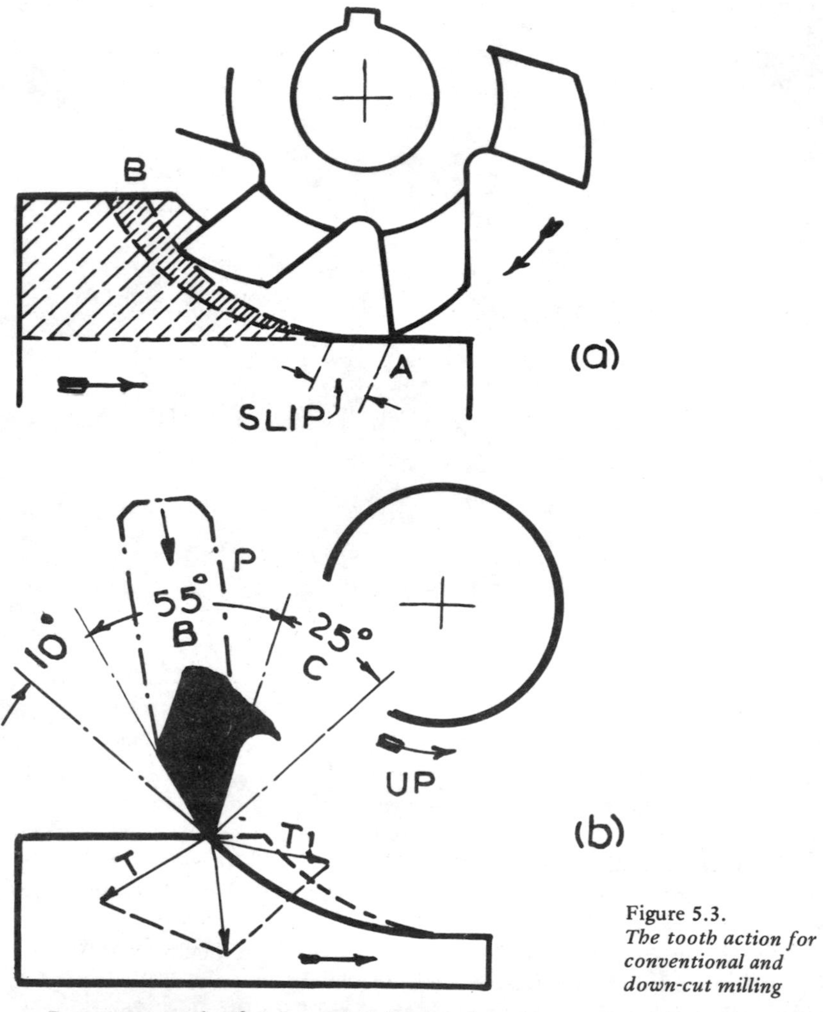

Figure 5.3.
The tooth action for conventional and down-cut milling

P represents the force needed for parting the metal, and, once cutting commences, the tooth cannot jump out as it is surrounded by metal on all sides while the forces *T* balance *P*. The cutter is driven by the peripheral force *UP*, but it is still directionally downwards. The angle *C* of the front face can be as much as 25° giving a small tool angle *B* of 55°.

BACKLASH ELIMINATION

On heavy milling machines elaborate means of eliminating backlash are used; one means is by using duplex feed screws for the table traverse, where the connection between the two is by helical gears which have a slight side movement to compensate for any wear between the nut and

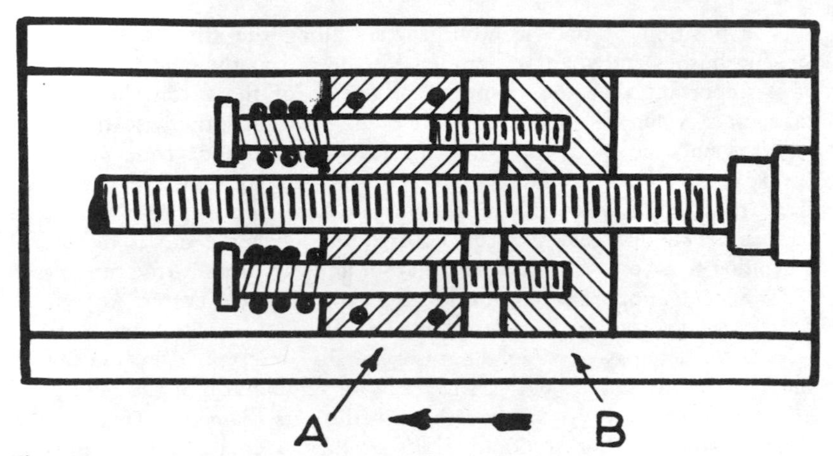

Figure 5.4.
A backlash eliminator to table traverse when milling:
A and B nuts

screws. Other methods employ hydraulic power for compensation, but a simple means suitable for lighter milling machines is shown in Figure 5.4. The two nuts A and B engage the feed screw; A is fixed to the table, while the second nut B is free to slide axially but is prevented from rotating with respect to the table as it is formed with a flat face which bears on the underside of the table. The nuts are forced together by the springs shown, thus taking up backlash in the feed mechanism when the table moves in the direction of the arrow for down-cut milling.

CUTTER MOUNTING

To locate the arbor in the spindle nose a standard taper of 3½ in/ft is used (Figure 5.5). This taper is too steep to be used for driving purposes and hence the use of the two keys on the spindle nose to drive the arbor. The end of the arbor is threaded so that a draw bolt can pass through the hollow spindle and can be used to pull the arbor into the taper bore and can prevent it from working loose during machining.

Figure 5.5.
An arrangement of the milling machine arbor and correct cutter mounting

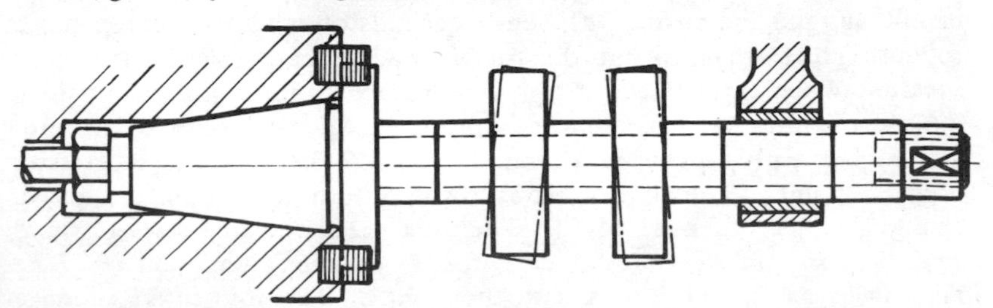

The position of one or more cutters along the arbor is secured by spacing bushes and then is clamped by the nut at the end of the arbor. It is important that the clamping or release of the cutter should only take place when the arbor support is in position; otherwise the arbor will certainly be bent, and the cutter will run out of true. Similarly, the spacing collars should be checked to see that they are clean and free from chips. The cutters shown in chain lines indicate what can happen when dirt gets between collars, for, when the nut is tightened, a bending stress can be set up on the arbor and can prevent true running.

The arbor support should be located as near to the cutters as possible to ensure rigidity, and similarly it is desirable to use an arbor with as large a diameter as possible. This is generally determined by the size of the hole in the cutter, but it should be remembered that the rigidity of an arbor increases by the fourth power of its diameter. Thus, if the diameter of an arbor is doubled, its rigidity becomes sixteen times as great.

It must be emphasized further the importance of keeping the distances between the arbor support, cutter and machine column as small as possible. The tendency to vibrate is a function of arbor deflection which varies to the third power of the length between the column and the overarm support. A milling machine arbor approximates a beam supported at the column end and the overarm bearing at the other. If W equals the load due to the chip pressure and L is the length between the supports, then the deflection of the arbor can be found from $\delta = WL^3/48EI$. Therefore δ increases as L^3 and inversely as d^4, where d is the diameter of the arbor. Thus assuming that d is 1 in (25.4 mm) and L is 12 in (305 mm) in one case, and 1¼ in (32 mm) and 6 in (152 mm) in another case, then the deflection of the first arbor will be twenty times as great as the second. It may be considered that a milling machine arbor approximates more nearly to a built-in beam than to a beam supported at both ends. In this case the deflection would be $WL^3/192EI$. In actual practice the deflection is between the two formulae given, and in any case the points mentioned previously are fully vindicated.

TYPES OF CUTTERS

The nature of the chip is responsible for most of the difficulties found in milling, and the cutting efficiency depends on whether a cutter tends to push rather than to cut the metal or whether the action is one of shearing. When cutters have straight teeth, each tooth begins to cut along its entire width, so that a shock is produced as each tooth comes into action, but with helical teeth as in Figure 5.6 the action is progressive with the point of contact moving across the tooth as rotation takes place. The angle of tooth is 25 to 30°. The usual clearance angles are for mild steel 6°, for hard steel 4°, for cast iron 7° and for aluminium 10°. The rake angle is 10°. The face A with the front of the tooth forms an angle

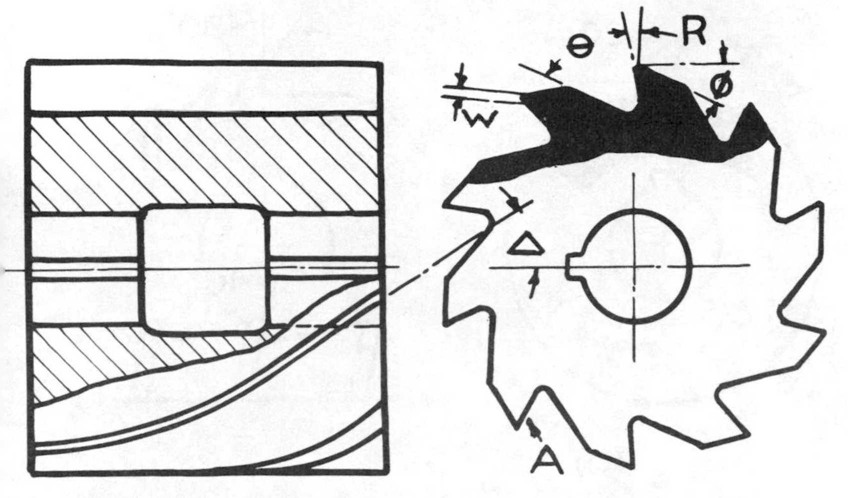

Figure 5.6.
A design for a spiral tooth milling cutter

which should be 50° irrespective of the diameter and pitch of the cutter. The spiral angle Δ is 25°.

For heavy slab milling, roller cutters as in Figure 5.7 give a good shearing action, for, as the full cut comes into operation, a spiral chip is produced by the two forces in action, i.e. a force at right angles to the tooth edge and a force parallel to the direction of the feed motion. This gives a resultant force in the direction of which the chip is sheared off, so that the direct thrust on the cutter is greatly reduced. The spiral angle is generally about 35° but may be made with an extra quick spiral up to 66° as shown. Examples include 3½ in (89 mm) in diameter and 6 in (152 mm) long with 3 teeth and a right-hand lead of 4.18 in (106 mm) or 4 in (102 mm) in diameter and 6 in (152 mm) long with 4 teeth and a right-hand lead of 6.25 in (158 mm). The end thrust might be thought to be excessive, but the amount is only about 6% of the total power for the cut.

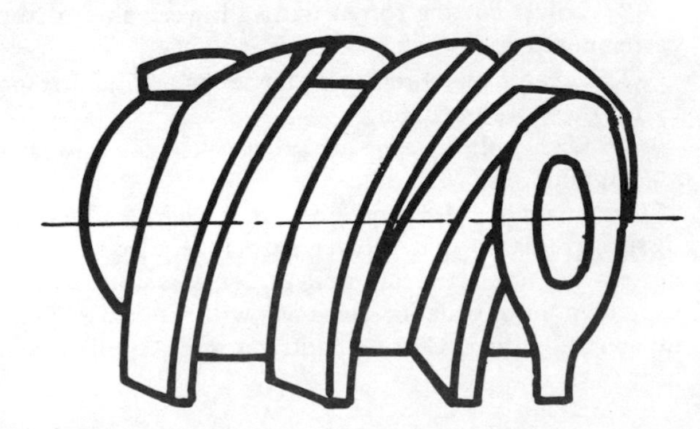

Figure 5.7.
A type of roller cutter for slab milling

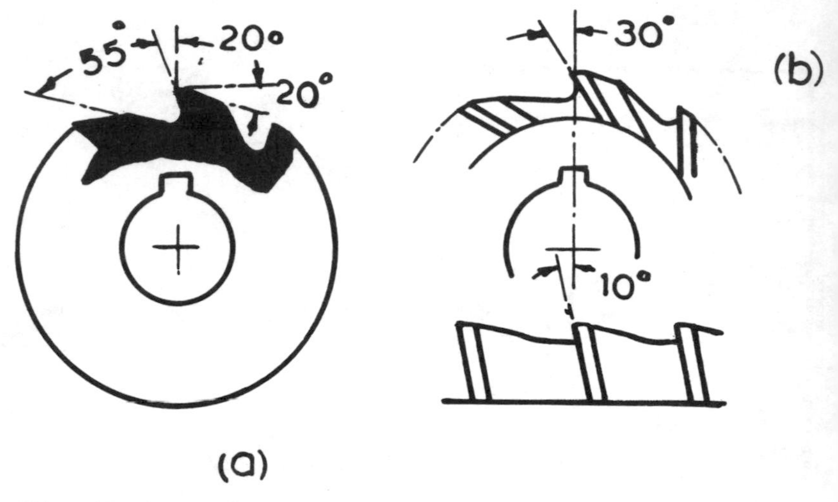

Figure 5.8.
Details of high-rake milling cutters

HIGH-RAKE CUTTERS

Milling cutters of high-speed steel are generally made with a rake angle varying from 5 to 15°, but it has been shown that metal removal can be higher and power consumption lower if rake angles are made much higher than the values given. Notwithstanding the acute edges formed by the higher rake, a considerably higher cutter life and operating speed can be obtained compared with those of conventional high-speed cutters.

A slab cutter for operating on mild steel and up to 60 ton tensile steel is shown in Figure 5.8(a). This has 9 teeth and a 35° right-hand spiral. The cutter shown in Figure 5.8(b) for side and face cutting has 14 teeth and a 10° helix angle. The high performance of these cutters is due to the following factors.

(1) A high shear angle and a correspondingly shorter path of shear.

(2) Lower cutting forces with a lower value of the chip–tool interface temperature.

(3) Greater metal removal efficiency and a correspondingly lower value for the work in cutting.

(4) Less built-up edge which results in less wear due to abrasion and chipping.

The main disadvantages are that there is less metal in the tooth to dissipate heat so that greater care is required when grinding. Also, these cutters do not stand up well under abrasion, so that it is better not to use them into scale or surfaces with abrasive inclusions, except when down-cut milling. Cutting fluid must be applied generously to dissipate heat.

Figure 5.9.
A vertical attachment for the Harrison milling machine

VERTICAL MILLING OPERATIONS

The spindle construction on a vertical milling machine is better able to resist cutting forces than with cutters mounted on an arbor, and the machines illustrated are available with vertical spindles. An alternative, however, and a useful attachment for any horizontal machine is a vertical milling head, as shown in Figure 5.9. This extends the use of the machine considerably, for it will swivel 45° either side of the vertical for angular work, and, when used in conjunction with a circular table, its use is still further increased. Normal work, apart from surfacing operations, includes the milling of tee slots, dovetails and angular faces of slideways.

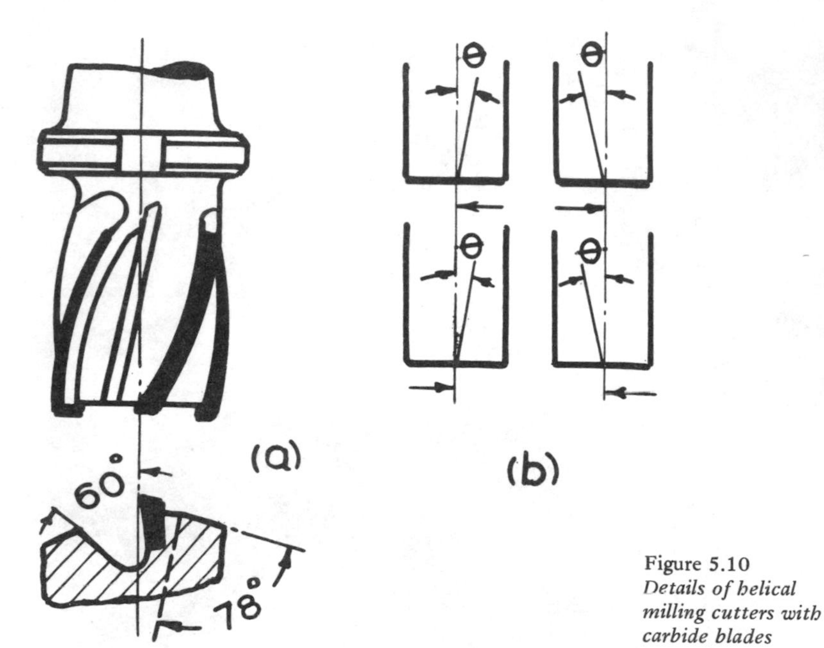

Figure 5.10
Details of helical milling cutters with carbide blades

HELICAL CARBIDE CUTTERS

Figure 5.10(a) shows one of these cutters designed by the Boeing Airplane Co. Helix angles from 15 to 20° are provided, the power required for a given cut decreasing with an increase in the helix angle. The main problem in manufacture was the twisting of the carbide blades to suit the slots in the body, but this problem has been solved by heating the blade to 2,300°F when it becomes plastic and can be twisted to shape. Cutter life has been increased 450% with a saving of 50% in sharpening time.

The six-bladed cutter shown has been used on alloy steel with a tensile strength of 200,000 lb/in^2 (140,000 kg/cm^2). Stock removal is at 200ft/min (60.8 m/min) with a feed of 10 in/min (254 mm/min). For the milling of aircraft spars in aluminium alloy, cutting speeds of 6,500 ft/min (2,000 m/min) are used with a feed of 280 in/min (7 m/min). The cutter used runs at 3,550 rev/min and has four blades. The front rake varies from 10 to 15° depending on the operation.

SPIRAL END MILLS

The designation for spiral tooth and rotation is indicated in Figure 5.10(b). Thus, left-hand rotation is clockwise, and right-hand rotation is anti-clockwise, looking on the front end of the machine spindle. Left-hand spiral is that in which the spiral rises from the left-hand side, and right-hand spiral is when the spiral rises from right-hand side. The combinations

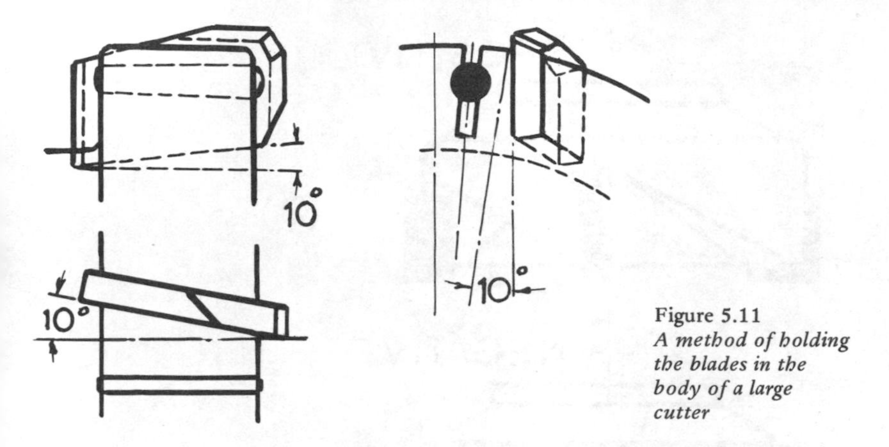

Figure 5.11
A method of holding the blades in the body of a large cutter

are important, giving in one case a positive angle to the front teeth, which tends to draw the cutter out of the spindle, and in the other case, a negative angle to the end teeth, which tends to push the cutter into the spindle.

INSERTED TOOTH CUTTERS

Large milling cutters would be too expensive if made of a cutting material, and so blades of a suitable cutting material are inserted into a cast iron or steel body. The blades may be of high-speed steel, but Stellite blades give better results, or the blades may be carbide. One method of holding is shown in Figure 5.11, where the angular setting in two directions simplifies the grinding of the blades for rake and clearance angles. To hold the blades in the body, saw cuts alternate with the blades, and the saw cuts are fitted with taper pins which when driven tight expand the metal to grip the blades.

MILLING WITH NEGATIVE RAKE

The limiting factor when milling with cemented carbide is the low tensile strength of the material, which is half the strength of high-speed steel. Figure 5.12(a) shows a cutter with a positive rake, and Figure 5.12(b) one with a negative rake. With a positive rake the forces tend to pull the tip away from its seating so that the cutting stresses must be taken by a brazed joint in tension; also such a tip must have an included angle of less than 90°. Tips of cemented carbide would fail under such conditions, but by using a negative rake angle of 10° the included angle at the tip is normally 90°, while the pressure of cutting tends to hold the tip against its seating. Another important feature is that the initial contact with the work is not made by the point of the tip but higher up, thereby protecting the edge and allowing a gradual build-up of load.

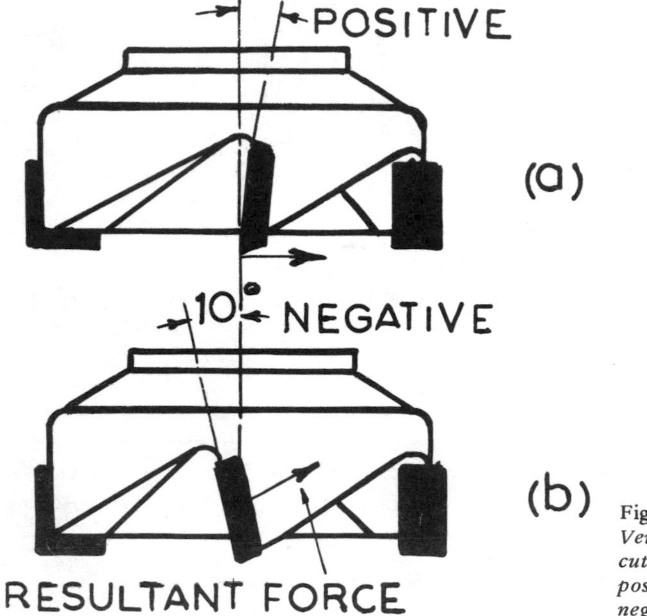

Figure 5.12.
Vertical milling cutters with a positive and a negative rake

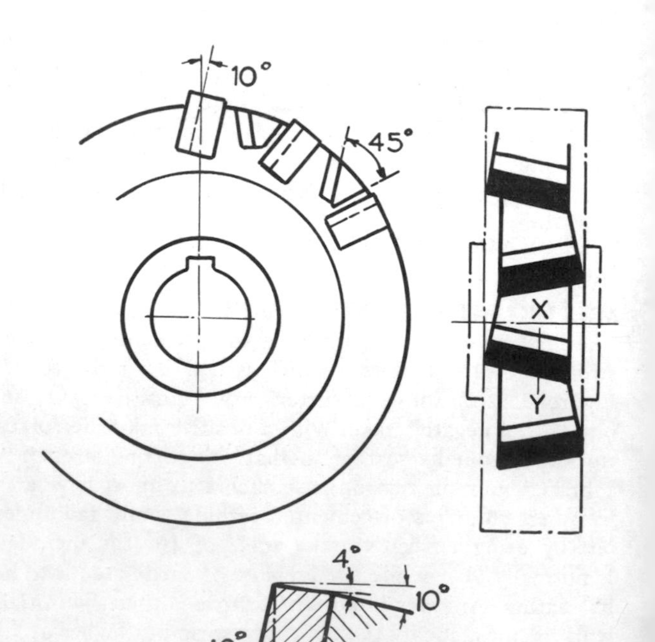

Figure 5.13.
A negative rake side and face milling cutter

Notwithstanding these improvements, successful operation can only be obtained at high speeds, for the cutting action is more like that of an ordinary cutter working backwards. Fortunately, the shear strength of a metal decreases as its temperature is raised, and by cutting at 600 ft/min (182 m/min) or more a temperature at which the metal can be machined can be achieved. The rapidity of cutting is such that the heat generated is largely dissipated with the chip and does not affect the work.

Figure 5.13 shows a negative rake side and face cutter with 16 teeth tipped with carbide and with 10° negative side and helical angles, 4° top rake and 0° side relief. It is 6 in (152 mm) in diameter and 18 in (35 mm) wide. Each tooth cuts like a lathe tool and is very effective in slot milling and for cutting large keyways, while an excellent surface finish is obtained.

Table 5.1.
Surface cutter speeds for milling

Material	Speed for high-speed steel		Speed for cemented carbide	
	(ft/min)	(m/min)	(ft/min)	(m/min)
Cast iron	75	23	600	184
Mild steel	75	23	600	184
0.4% carbon steel	50	15	400	122
Bronzes	120	36	1,000	302
Aluminium	150	45	2,000	604

The feed per tooth is about 0.006 in (0.15 mm) for machining steel or

$$\text{feed per tooth} = \frac{\text{Table traverse}}{(\text{rev/min}) \times \text{no. of teeth}}$$

SHARPENING OF MILLING CUTTERS

Milling cutters may be divided into two general classes.

(1) Straight or spiral tooth cutters.

(2) Form-relieved cutters.

Figure 5.14(a) shows the clearance angles for solid milling cutters of type (1), in which the primary clearance A is located behind the cutting edge, and the width of this clearance forms the land which should range from $\frac{1}{32}$ to $\frac{1}{16}$ in (0.8 mm to 1.59 mm) in width. Following this is the secondary or backing-off clearance B which is ground to relieve the first clearance. After regrinding to D the width of the land is increased and restored to its original width. The rake angle is C while the shaded portions indicate how the original width X is maintained at X' by correct grinding. Figure 5.14(b) shows a form cutter in which grinding can only be carried out on the front face as shown; otherwise the form would be changed.

Table 5.2.
The angles for milling cutters for high-speed steel

Material	Primary clearance (in)	Secondary clearance (in)	Rake angle (deg)
Mild steel	3 to 4	6 to 10	10 to 15
Alloy steel	2 to 3	6 to 10	10 to 15
Cast iron	3 to 4	6 to 10	8 to 10
Brass, bronze	2 to 4	6 to 10	10 to 15
Aluminium	6 to 10	6 to 10	10 to 35

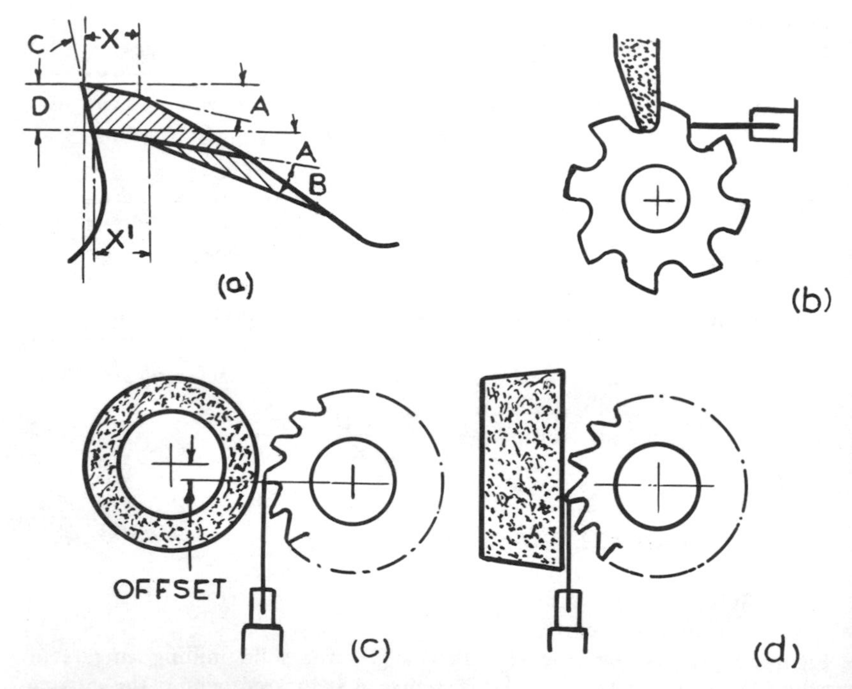

Figure 5.14.
Methods of grinding teeth on milling cutters

GRINDING WHEELS

For cutter sharpening, wheels should be of a medium-soft grade and not too fine, for fine wheels cut slowly and tend to burn the cutter teeth. If the wheel is too soft, it will wear rapidly, which may affect the diameter of the cutter that is being ground unless a wide wheel is used. The advantage of a wide soft wheel is the elimination of the danger of burning. Cutters may be sharpened by using either the periphery of a plain wheel (Figure 5.14(c)) or the face of a cup wheel (Figure 5.14(d)). The latter will grind a flat land, whereas the peripheral wheel leaves the teeth slightly concave which results in more rapid dullness than if the land were flat.

Excessive clearance will cause chatter and will dull the teeth. For clearance on end teeth on end mills 2° is suitable. The clearance angle is regulated by setting the centre of the wheel slightly above the cutter or by adjusting the tooth rest below the centre as shown. The safe way when grinding is to revolve the wheel so that the cutting action keeps the tooth down on the work rest. The grit and sparks then fly downwards, and the necessity of holding the cutter down on the work rest does not arise.

CEMENTED CARBIDE CUTTERS

If the teeth of a cutter are only slightly worn, then a diamond wheel can be used for grinding and backing-off, but normally the first grinding operation is by the use of a silicon carbide wheel. Following this, to bring the cutter to the required condition the teeth must be diamond lapped, and the cutting edge must be lightly honed. Diamond wheels and laps should be dressed and trued only as occasion demands, and not as frequently as ordinary vitrified wheels; otherwise too much diamond which is expensive is lost. For dish or cup wheels, they should be removed from the mount and should be rubbed by hand in a circular motion on a sheet of glass which is covered with 100 grit carborundum mixed with water to a thin paste. Coolants are recommended with resinoid-bonded wheels, but only a small amount is required if using light lubricating oil, but the wheel must not run dry.

SAFETY PRECAUTIONS

To comply with factory regulations, all milling machines must carry a guard which should enclose as much of the cutter as possible (see BS 5304). Vertical milling machines have always presented the greatest problem, but this has been solved by Vickers patent guard shown in Figure 5.15. This is sold as the Max-A-Just-guard (The Silvaflame Co. Ltd) and has the advantage in that the sections are held in tee slots of the machine table and are adjustable in both horizontal and vertical directions. The acrylic extension visors enable the operator to have a clear view of the cutter during machining, while the release of a locking device enables the assemblies to be opened to give rapid access to the workpiece. Where tee slots are not available, guards are fitted with magnetic bases.

Accidents on milling machines are usually caused by one of the following.

Figure 5.15.
An efficient means of guarding milling cutters
(By courtesy of Silvaflame Co. Ltd, Walsall)

(a) Attempting to remove swarf by using a spanner, steel rule or similar implement near the cutter in motion.

(b) Adjusting pipes for coolant in the vicinity of a moving cutter.

(c) Using rags for cleaning parts of a machine near cutters, and on occasions the trapping of swarf brushes or loose clothing.

(d) Removing the work from a vice or fixture which has not been withdrawn to a safe distance from the cutter.

(e) Guards not adequate to prevent cuttings from flying and causing eye injuries, particularly when machining light metals at high speeds.

(f) Attempting to measure the workpiece with the cutter running and with the feed motion in operation.

(g) Failing to see that all clamps are tight before commencing cuttings.

The modern milling machine is not dangerous in itself, as all the mechanism is adequately enclosed and protected, and the observance of the recommendations (a) to (g) will ensure safety of operation.

MILLING OPERATIONS

Figure 5.16(a) shows an example of plain slab milling, the work being held in a vice. This is the simplest milling operation and the simplest means of holding work. A vice with a swivel base, as shown, extends its usefulness in that grooves, for example, can be milled at any angle to each other without resetting the work. Figure 5.16(b) shows the

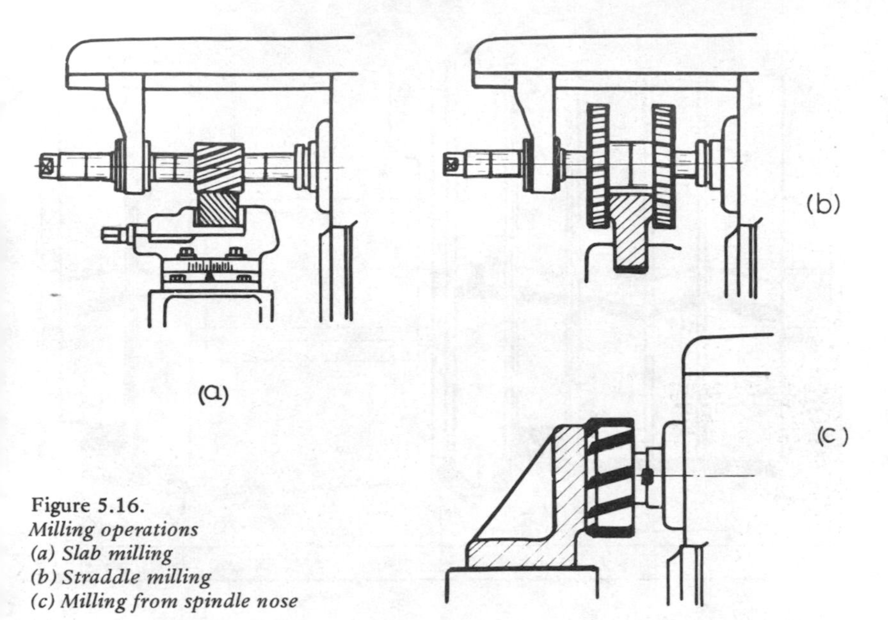

Figure 5.16.
Milling operations
(a) Slab milling
(b) Straddle milling
(c) Milling from spindle nose

operation of straddle milling in which case the sides of the teeth are used instead of cutting on the periphery. It should be noted that the arbor support has been located close to the cutters. Figure 5.16(c) shows the use of a cutter when milling from the spindle nose. Face milling can be carried out by attaching the cutter to the spindle flange. The work is set vertically, and the depth of cut is adjusted by the cross-traverse handwheel. The knee slide and cross slide should then be locked for milling to take place. For other operations such as grooves or tee slots, an end mill would be fitted into the bore of the spindle nose and would be held by the draw bolt.

PENDULUM MILLING

If several cutters are mounted together, the operation is known as gang milling. If cutters of helical teeth are used, it is advisable to balance the end pressure by using cutters with right- and left-hand teeth and to mount them on the arbor so that they cancel out any end thrust. Figure 5.17 shows a gang of cutters used for milling lathe saddles by the system of pendulum milling; the work is mounted in a fixture at each end of the milling table. After a component in one fixture has been machined, the cutters return to mill the other one.

The problem is that normal milling is taking place in one direction and down-cut in the other, so that the optimum conditions for efficient machining cannot be obtained in both directions. Nevertheless, by using suitable cutters very high production can be obtained for fixture loading

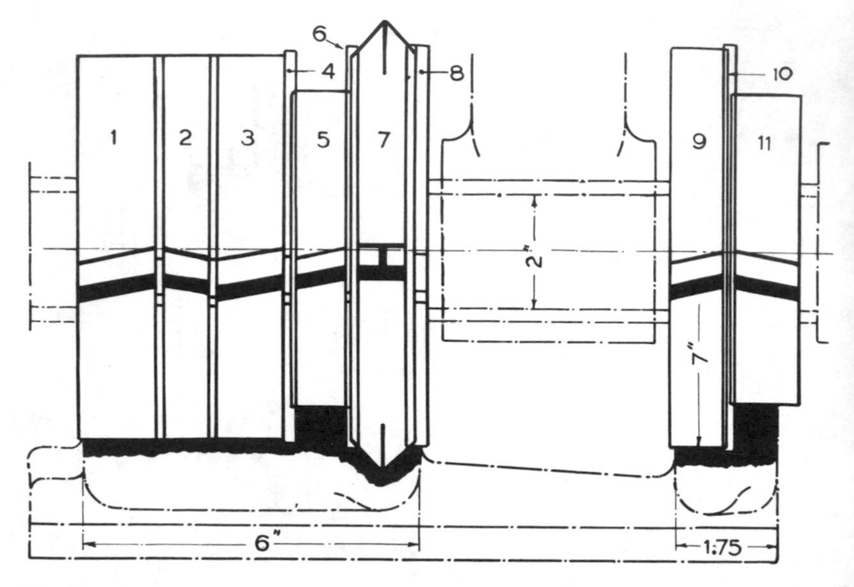

Figure 5.17.
A set-up for pendulum gang milling for the saddle of the Harrison lathe

at one end of the table which is proceeding while milling is taking place on the other one. All 11 cutters are carbide tipped with 6 teeth of 5° radial rake, and with the exception of numbers 4, 6, 7, 8 and 10, with the teeth at an angle of 10° straight across the face. This gives better results when milling than spiral teeth on down-cut operations, but the face width is limited; so to cover the broad face on the left-hand side of the work, three interlocked cutters of 172 mm are used.

All cutters have primary angles of 5 and 7° secondary, while diameters range from 127 mm for number 5, to 200 mm for the vee cutter. Speeds must be based on the largest diameter, and this is 60 m/min with a feed of 140 mm/min and a depth of cut averaging 5 mm. An important feature is that, while two fixtures are required, only one set of cutters is used, as against two sets for equivalent work on two machines. Also, the loading and unloading time is not charged to the operation, for the machine is working productively at the same time.

The cycle time can be estimated from Figure 5.18 with the cutter shown in the stopped position in the first diagram and half the cycle indicated in the remainder. The same cycle is then repeated for milling at the opposite end of the table. To obtain the time to produce one component, A, O_1, O_2, L_S and L_A represent the approach, the overtravel, the length of cut and the minimum safe loading distance from the cutter.

Then

$$t_m = \frac{A + O_1 + O_2 + L_S}{F} = \text{milling time per piece}$$

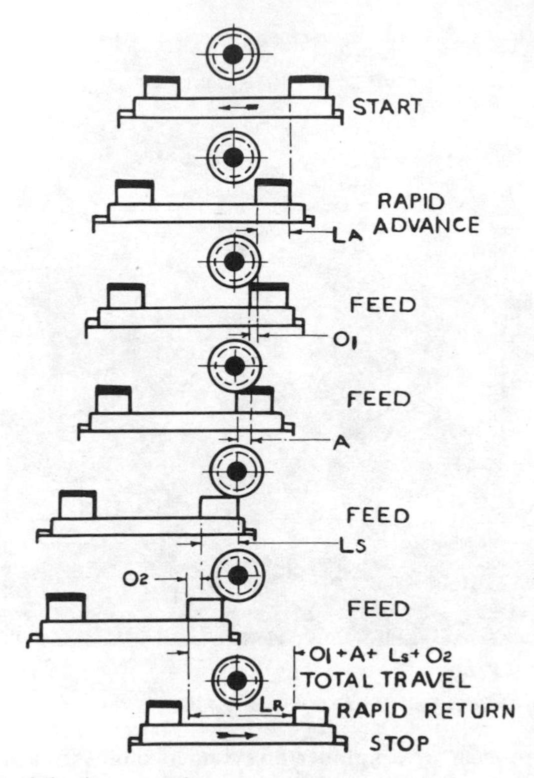

Figure 5.18. *Diagrams showing the time cycle for pendulum milling*

and the idle time per piece

$$t_i = \frac{A + O_1 + O_2 + 2L_A + L_S}{R}$$

thus the cycle time per piece is

$$t_c = t_m + t_i + t_6$$

This figure t_6 can be ignored as a rule, for it merely represents the time to start the cycle, and in most cases there is no necessity to stop the machine unless it is necessary to clear away cuttings.

WORK HOLDING WHEN MILLING

As the operation is an intermittent cutting process, it tends to produce vibrations more readily than when cutting with a single-point tool. Thus the work must be well supported and clamped to avoid any tendency to vibrate or spring which may affect the dimensional accuracy. This is particularly the case when gang milling with many cutters as just described. Actually, the bed of the lathe on which the saddle fits requires a set-up of 14 cutters for the slideway tops, while the underside is milled by the cutters shown in Figure 5.17. The interesting feature is the hydraulic clamping of the beds by hand-operated units (Figure 5.19). These operate at pressures up to 1,500 lb/in^2 (105 kg/cm^2) and work from a hand pump

Figure 5.19.
The hydraulic clamping of Harrison lathe beds for gang milling

which supplies oil to each of a number of cylinders built into a fixture. This system is used for holding two and sometimes four beds. The pump is shown at the bottom centre, and pressure is applied by the capstan wheel; all clamps are operated simultaneously and are released by spinning back the wheel. A pressure gauge is fitted to indicate holding pressure. A second hydraulic unit for clamping another pair of beds while machining is proceeding on the first pair is shown at the bottom left-hand corner of the illustration (Power Jacks Ltd).

ATTACHMENTS FOR THE MILLING MACHINE

The range of operations which can be performed on the milling machine is considerably extended by the use of suitable accessories and difficult operations can be simplified. Thus output of work is increased. The use of a vertical milling attachment on a horizontal machine has been described, but smaller equipment includes machine vices, plain or swivel type, angle plates, tilting tables and vee blocks. Another useful accessory is a slotting attachment indicated in Figure 5.20(a) which can be used for cutting keyways, square or hexagon holes, or indexing. As the attachment can swivel through 360°, work can be slotted in any plane and, when used in conjunction with a circular table, can be used for the graduation or the production of dies.

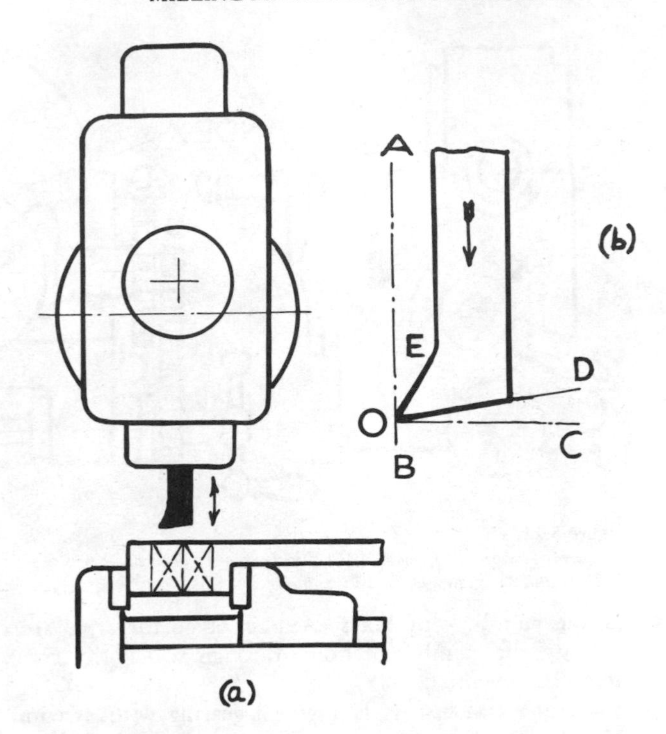

Figure 5.20.
A slotting attachment for a milling machine

As the attachment is driven directly from the main spindle of the machine, the speeds of the spindle represent the number of working strokes per minute of the tool slide. The example shows the slotting of a square hole in a lever for which it is necessary for a hole to be drilled previously to slotting. After completion of one edge of the square, the tool can be adjusted in the holder, or the work can be indexed on a rotary table. Figure 5.20(b) shows the shape and action of a slotting tool. The traverse of the tool along the line AB changes the position of the tool angles compared with a shaping tool. If *O* is the tool point and *O*C is normal to the work surface, then the angle C*O*D forms the front rake and A*O*E the relief. A suitable cutting speed on mild steel and cast iron is 100 ft/min (30.4 m/min).

DIVIDING HEAD

The provision of an indexing or dividing head with another unit, i.e. a small tailstock, whether used for simple indexing or the more complex operation of universal milling is very useful. As shown in Figure 5.21, the indexing to obtain any number of divisions is effected by turning a crank carrying a locating plunger past an appropriate number of holes on a pitch circle of equally spaced holes. The basis of the movement is the index plate E, which has several concentric pitch circles with

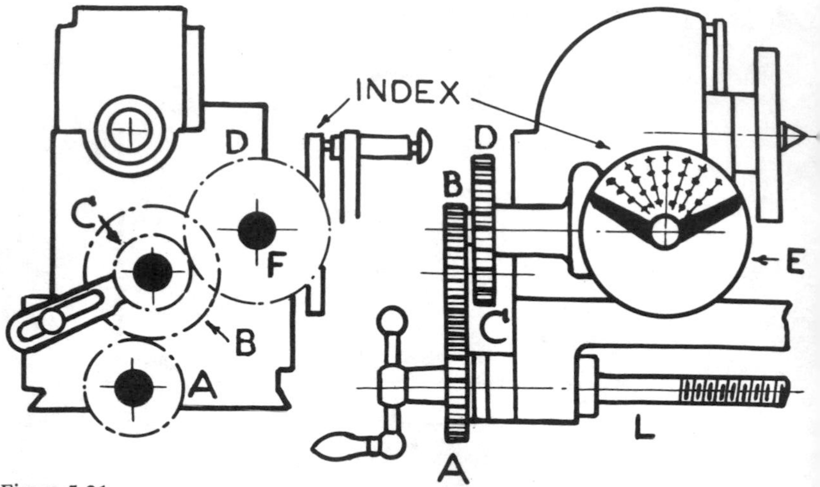

Figure 5.21.
The arrangement of gearing on a dividing head for spiral milling:
A, B, C and D gears E index plate

various numbers of holes. A pair of sector arms are clamped so as to indicate the number of holes through which the crank must be turned at each movement.

On the crank shaft is a worm gearing with a worm wheel which has 40 teeth, so that 40 turns of the crank produces one revolution of the spindle. Thus to obtain, say, 13 divisions the number of turns of the index crank would be $40/13 = 3\times(1/13)$ turns, or by using a plate with 39 holes it would be $3\times(3/39)$.

ANGULAR INDEXING

One turn of the crank is $360/40 = 9^\circ$; therefore one ninth of a turn is 1°. As an example, to index 35°, the number of turns of the crank is $35/9 = 3\times(8/9) = 3$ turns and 8° (16 holes in an 18-hole circle gives 8°). To index to minutes, one turn of the crank is $9 \times 60 = 540$. For example, to index $16'$ requires 1/34 turn because $540/16 = 34$ approximately. As the 33 hole is nearest, it would be used with a very small error.

SPIRAL MILLING

By connecting the dividing head to the table screw L the spindle can be caused to rotate as the table moves along, and by this means spiral gears, helical flutes and splines can be cut.

If gears of equal size are employed so that the screw L and shaft F rotate at the same speed and if screw L has four threads to the inch, then the table movement will equal $40/4 = 10$ in. Thus, to mill a spiral with a lead of 12 in, the gear ratio would be 12/10 or the lead of the spiral divided by the lead of the machine. Suitable gears would then be selected to give this ratio and mounted (A/B) × (C/D) as shown. It is then required to know the spiral angle so that the table can be swivelled to suit the angle. This can be found from

$$\text{tangent of angle} = \frac{\text{circumference of work}}{\text{lead of spiral}}$$

For example to mill a spiral with a lead of 48 in on a workpiece with a circumference of 12 in, then $12/48 = 0.25 = \tan 14^\circ$; therefore the table would be swivelled to this angle.

Figure 5.22.
A modern milling machine of bed-type construction; the vertical head may be removed to enable normal horizontal milling to be undertaken. Either NC or manual control may be used, the latter functions being selected from the pendant. Hardened slideways and recirculating ballscrews are a feature of this machine together with a swarf conveyor seen on the left
(By courtesy of Oerlikon Italiana S.p.A.)

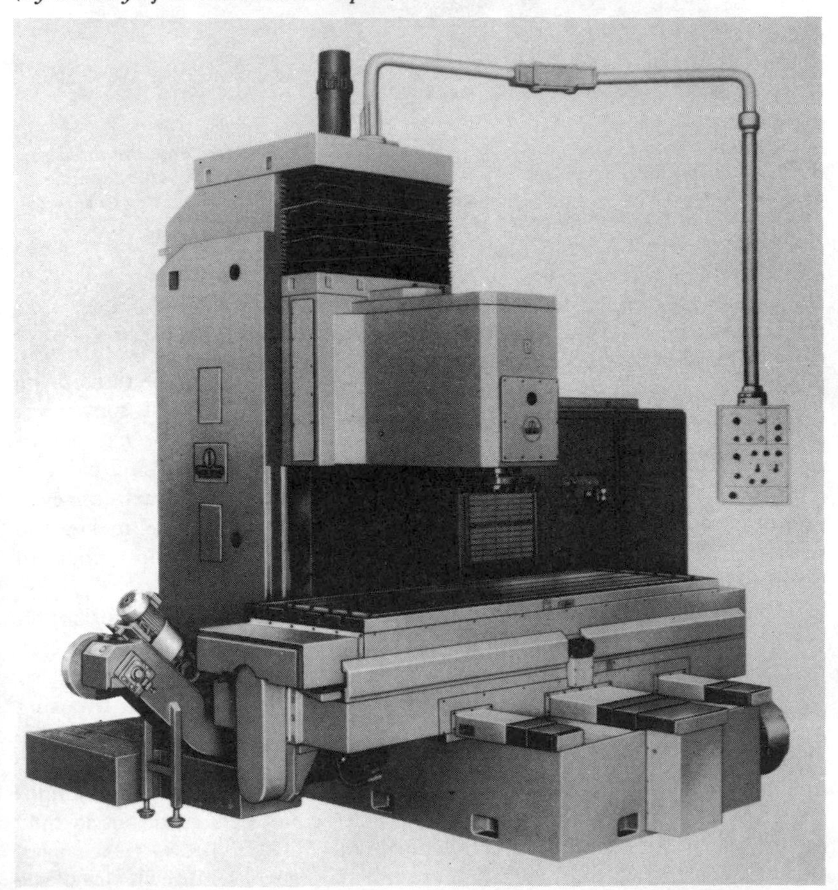

Figure 5.23.
A fully universal head equipped with its own motor drive and gear box, mounted on a horizontal machine. The head may also be driven from the main machine spindle by using a suitable mounting plate. Compound angles may be milled using this device (By courtesy of Oerlikon Italiana S.p.A.)

BED-TYPE CONSTRUCTION

A milling machine (Oerlikon Italiana S.p.A.) for heavy duty is shown in Figure 5.22. Either NC or manual can be used, the latter functions being selected from the pendant which can be swung to the most convenient position for the operator. This pendant control reduces the number of operating levers to a minimum. Hardened guideways and recirculating ballscrews are a feature of the machine, together with a swarf conveyor seen on the left. The high metal removal of this machine makes the problem of swarf removal serious unless a mechanism is provided to transport the swarf to a convenient receptacle or conveyor belt.

Figure 5.23 shows a fully universal head mounted on a horizontal machine. The head is equipped with its motor drive and gear box so that the machine is converted for vertical operation. The head may also be driven from the main spindle, shown below the attachment, by using a suitable mounting plate. Twelve spindle speeds are available and are operated by the levers on the side of the slide.

Figure 5.24 shows that the spindle can be swivelled to any angle, simple or compound. This is accomplished by a three-piece construction so that the head can be swivelled around the main slideway for vertical angles and can then be further swivelled around the centre section in the other

Figure 5.24.
The same universal head as shown in Figure 5.23. but adjusted for machining a compound angle. It should be noted that the machine is equipped with an optical reader on the table for accurate positioning; the cross movement of the knee is similarly equipped

plane, thus compounding the angles to any position and making the head fully universal for machining purposes. It should be noted that the machine is equipped with an optical reader on the table for accurate positioning, the cross-traverse table movement being similarly equipped.

6

THE GRINDING PROCESS AND OTHER FORMS OF ABRASIVE MACHINING

TYPES OF ABRASIVE MACHINING

Grinding is the most common form of abrasive machining; it demands some means of moving a rotating grinding wheel relative to the workpiece. Descriptions of the various types of grinding machines have been given in Chapter 2, and therefore this chapter is concerned only with abrasive materials, the wheels used for grinding and certain other abrasive purposes. The most common uses of abrasive machining are where hard material is found which cannot be readily cut in any other manner, where high accuracy is needed or where high degrees of surface finish are essential. Abrasive machining should also include, in consequence, lapping, honing, superfinishing, polishing and deburring.

ABRASIVES

Natural abrasives such as emery and corundum were used for a long time in the early days of engineering, but, whilst emery is a tough and durable abrasive, it contains iron and other non-cutting elements which reduce its efficiency. Corundum, an oxide of aluminium like the sapphire and ruby is purer than emery and contains a larger proportion of crystalline alumina which is the cutting element in both abrasives. For a number of years, however, more reliable and efficient abrasives have been produced in the electric furnace; these are silicon carbide and aluminium oxide. Carborundum and Crystolen are well-known trade names for the former, whilst the latter is frequently referred to as Aloxite or Alundum.

Silicon carbide is manufactured by the chemical interaction of sand and carbon, crystals being grown in the furnace at temperatures around 5000°F. A typical process may take around 35 h, and after cooling the sharp crystals are removed and crushed prior to further processing.

Aluminous abrasives are made in an electric arc furnace by the fusion of bauxite.

BONDING

Apart from the abrasive, a common grinding wheel has a second major ingredient which cements the abrasive grains together and is called the bond. The bond must hold the grains firmly and yet must have an open structure so as to permit the abrasive to cut freely, whilst in very thin wheels it must also have a certain amount of elasticity to ensure against wheel breakage. A wide range of bonds is available to suit various working conditions and which include vitrified glass or porcelain, sodium silicate, shellac, rubber and a variety of synthetic resins.

Wheels with a vitrified bond are frequently used and are made by mixing the abrasive grains with ceramic materials such as clay, feldspar or flint, by moulding the mixture to shape and by firing it in a kiln. After removal from the kiln the wheels are trued in a lathe, usually by using diamond-tipped tools, and are prepared for mounting by bushing the hole with lead or a plastics material. An overspeed test is made on each wheel to ensure that it is sound and will not break in service.

Some wheels after moulding and firing are fixed to a metal plate or cup to give additional strength for special applications. For finishing the very hardest of materials such as cemented carbide tools, diamond dust is bonded with metal or synthetic resins and is completed by a metal backing plate.

BORON ABRASIVES

Experiments are being carried out to popularize grinding wheels which incorporate boron carbide as a less expensive alternative to diamond wheels, and boron carbide hand laps are used on cemented carbide tools as well as for general industrial lapping purposes.

BORAZON

Borazon is the trade name of the General Electric Co., U.S.A., for cubic boron nitride, a synthetic by-product of research that created man-made diamonds. It is the second hardest substance known with a value of 4,700 on the Knoop scale as against that of aluminium oxide which is 2,100 and that of silicon carbide which is 2,480, the diamond still leading with a value of 7,000. Although wheels are available for both external and internal grinding operations, at the present time they are very expensive and are finding limited applications for grinding of 'difficult' materials such as hardened steels.

SELECTION OF GRINDING WHEELS

The grain or grit size of a wheel is designated by a number, say, 36, which is chosen because the grit will pass through a mesh of 36 pitches in the linear inch; the percentage of smaller grains is also controlled. Another term grade refers to the tenacity with which the bond holds the grains and does not refer to the hardness of the actual abrasive. The grade is coded by letters which are referred to later in this chapter. Whilst in use, two other terms are used to describe the behaviour of grinding wheels: glazing and loading. A glazed wheel has suffered dulling or wearing of the sharp points of the abrasive which causes slow cutting and generates excessive heat; a loaded wheel has particles of metal adhering to the cutting surface, filling the pores and severely reducing its cutting ability. In general, silicon carbide is recommended for grinding materials of low tensile strength, whilst aluminium oxide should give the best results on materials of high tensile strength.

Provided that the selection of grinding wheel has been relatively close to the optimum for the intended operation, the machining parameters can be adjusted to make the wheel act in a harder or softer manner or indeed, to strike a happy medium between the two. If a wheel glazes and cuts slowly, it is too hard, whereas, if a wheel wears quickly and loses its shape or roundness, it is said to be too soft. To make a wheel behave harder, the following procedure can be adopted.

(1) Reduce the work speed and keep the wheel speed constant or increase the wheel speed and keep the work speed constant.

(2) Decrease the traverse rate.

(3) Decrease the in-feed.

To make a wheel act softer, the following is applicable.

(1) Increase the work speed and keep the wheel speed constant or decrease the wheel speed and keep the work speed constant.

(2) Increase the traverse rate.

(3) Increase the in-feed. The greater the surface contact between the wheel and work, the softer the wheel should be.

STANDARDIZED MARKINGS

A standartized marking system is employed which has been devised by the Abrasive Industries Association to embody the essential information for both user and manufacturer when ordering grinding wheels. This simple system enables wheels to be duplicated exactly when re-ordering, but it should be emphasized that some difference can be experience between wheels of the same coding but from different manufacturers, as there are considerable difficulties in measuring the physical properties of bonded abrasive products in terms of their actual grinding action.

The system is flexible, allowance being made for the inclusion of any optional symbols, alphabetical, numerical or both, at the discretion of the manufacturers. The elements of the British Standards marking system are shown in Figure 6.1. From this it will be seen that the four main markings are as follows.

(1) Abrisive.

(2) Grit size.

(3) Grade.

(4) Bond type.

Optional markings can be found in the prefix to the abrasive code to denote the manufacturers' abrasive-type symbol and the inclusion of structure if desired; equally, a suffix may be added after the bond code to denote the manufacturers' wheel type.

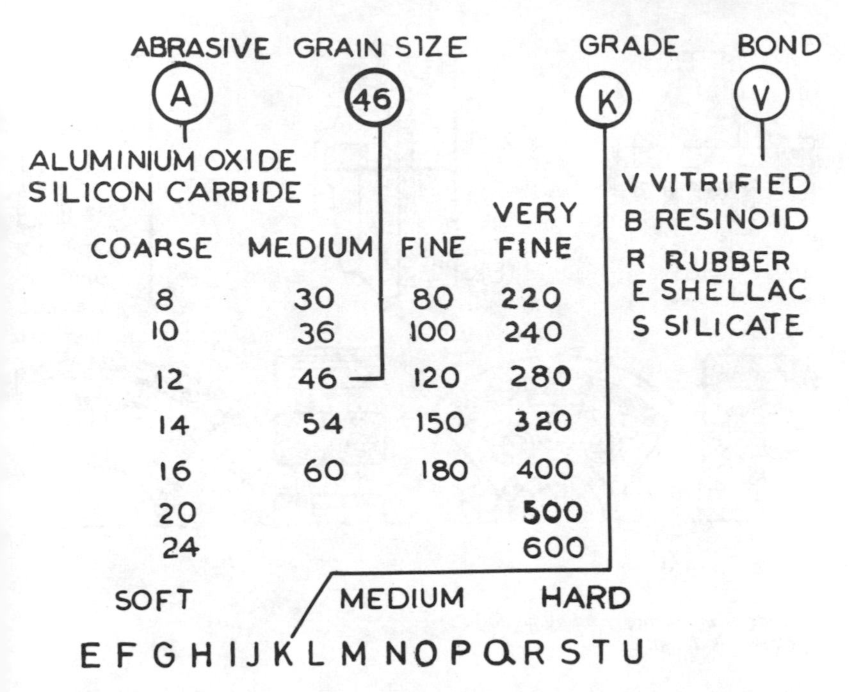

Figure 6.1.
Standard marking systems for grinding wheels

WHEEL MOUNTINGS AND WHEEL SHAPES

Grinding wheels are made in many different shapes and sizes in addition to the grit and bond variations already mentioned, but in all cases correct mounting of the wheel is of paramount importance.

Incorrect operation caused by unbalance and possible breakage of the wheel are the results of incorrect mounting. Plain cylindrical wheels are the most common type, and, as shown in Figure 6.2(a), a wheel should be held between flanges, preferably of steel, of equal diameter with the inner flange keyed or otherwise held on the spindle and running true with it. Normally, flanges should not be less than one-third and preferably up to one-half of the wheel diameter. The flanges should grip the wheel in the area around their perimeter rather than their inner diameter and must be recessed as shown at A. Washers of rubber, blotting paper or resilient plastics must be fitted between the flanges and the wheel to prevent damage to the wheel whilst gripping it with sufficient pressure for the grinding operation. The wheel should be an easy sliding fit on the spindle whilst not exhibiting any slackness, and the retaining nut should have a thread whose helix is in that direction which will tighten the nut with the rotation of the spindle.

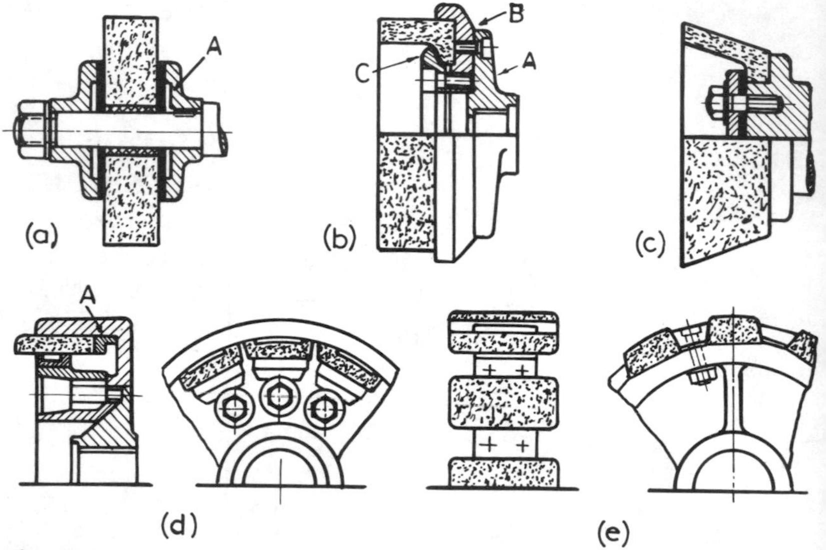

Figure 6.2.
Methods of mounting grinding wheels:
A plate B and C rings

When mounting a wheel of the ring or cup type (Figure 6.2(b)), it should be noted that the plate A is usually a fixture on the machine spindle, whilst rings B and C form a collet held together by screws, the whole assembly then being attached to plate A by more screws.

Ring B adds great strength to the wheel against bursting, and to mount a wheel it should be heated until flake shellac in the groove is melted. A slight turn is given to the wheel, as it is placed in position to ensure even distribution of the shellac and then it is left to cool; the wheel itself should be slightly warmed prior to this operation to prevent a sudden chilling of the shellac. When cold, the ring C is tightened in position with a Neoprene, rubber or soft leather gasket inserted between the wheel and the ring.

To avoid the heating process, Portland cement may be used instead of shellac, but 60 hs must be allowed for it to dry completely. Taper cup rings may be mounted as shown in Figure 6.2(c). The wheel is an easy fit on the spindle of the machine and is held against the flange by a plate and screws. As in previous examples an absorbent ring is placed between the plate and the inside of the wheel; care must be taken to see that the ring is clear of the inside radius of the wheel, as otherwise breakage may occur.

The development of segmental grinding wheels was initially prompted by difficulties in the manufacture of large solid wheels, but in recent years the development of high-powered grinding machines capable of removing metal at a rate comparable with milling machines has caused much research and improvement of segmental wheels. This is particularly true as wheels of large diameters and thicknesses up to more than 600 mm are in common use in some industries, and many manufacturing problems have had to be overcome. Segmental wheels offer certain advantages where they can be applied; damage to a single segment in use means its replacement and not the replacement of a complete wheel; a complete set of segments can be fitted at a much lower cost than that of a wheel, whilst the segments can be made deeper in section than a solid face grinding wheel. Also it can be shown that the effect of the segments in operation is beneficial to heavy grinding operations as the spaces between the segments permit coolant to enter readily, thus cooling the wheel and the work sufficiently, and the large flow of coolant also removes cuttings and loose abrasive readily.

A method of ensuring solid contact between clamps and segments is shown in Figure 6.2(d). The segments have a clearance in the curve at the back to prevent contact in the centre only. Steel or brass wedges are used to hold the segments, and as wear takes place, a filling-in ring A may be fitted. By the use of several different depths of rings, as much as 175 mm of a 200 mm deep segmental wheel may be used. In this type of design, typically, 12 segments are used in a chuck 800 mm in diameter and 15 in a chuck 900 mm in diameter.

Segments may be attached to the outside of a chuck in the manner shown in Figure 6.2(e).

INTERNAL GRINDING AND FACING

Difficulty can often be experienced in machining the face of a component flat and square to the bore. Previously turned workpieces can be ground to achieve this by mounting a face grinding wheel on the normal internal grinding spindle, but a better method is to have a separate spindle for each operation. This arrangement has been adopted by most manufacturers of grinding machines, and four typical short-batch examples of this type of work are shown in Figure 6.3. The machine used in each case was a Newall grinding machine.

Figure 6.3(a) shows a nickel-chrome bearing sleeve ground on the surfaces indicated. Nominal stock removal is 0.17 mm, and the bore must be within 0.012 mm of nominal size with a wall thickness within 0.005 mm. All diameters must be concentric and round within 0.001 mm with a high-grade finish. The time per piece is 10 mins.

Figure 6.3(b) shows a road breaker part which is made from a mild steel stamping and is required to have a good commercial finish. The tolerance on the bore diameter is 0.012 mm and on the outside diameter it is 0.007 mm. The stock removal (nominal) is 0.17 mm on both bore and outside diameter, whilst required roundness and concentricity is within 0.005 mm. The time per piece is 4½ min.

Figure 6.3(c) which requires internal grinding only is of interest in that

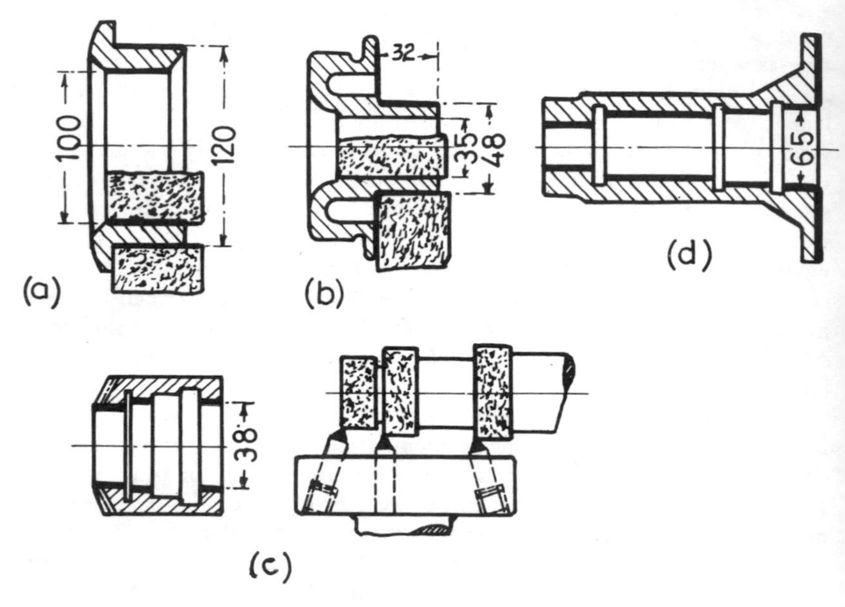

Figure 6.3.
Examples of internal and external grinding operations
(By courtesy of Keighley Grinders Ltd, Keighley)

three internal grinding wheels are mounted on the one spindle and are dressed simultaneously by three diamonds held in a special holder. This arrangement permits all three bores of a chuck sleeve to be ground simultaneously. The material is alloy steel, and all diameters must be round and concentric within 0.001 mm. Stock removal is 0.22 mm, and a high-grade finish is essential. Grinding time is 3 min per piece.

The fourth example (Figure 6.3(d)), that of a rock drill body, is a mild steel stamping which requires the grinding of four bores and the flange. The stock removal is nominally 0.265 mm, and a good commercial finish is required. The bores must be round and concentric within 0.005 mm, and the time per piece is 30 min. By using the method of internal and face grinding on separate machines, the time required was 2½ h, clear evidence of cost-saving advantages to be made whilst meeting accuracy requirements.

Figure 6.4 shows six high-volume production examples from the automotive and bearing industries. The first three (Figure 6.4, parts 1, 2 and 3) show interesting examples of the manner in which internal and external grinding are tackled, where the grinding machine can be designed around the component. All the grinding on each component is undertaken simultaneously, which results in a further saving in time and cost, and it will be noted that the external wheel approached the work from an angle, thus permitting rapid grinding of a face with the periphery of the wheel whilst also grinding a diameter with it.

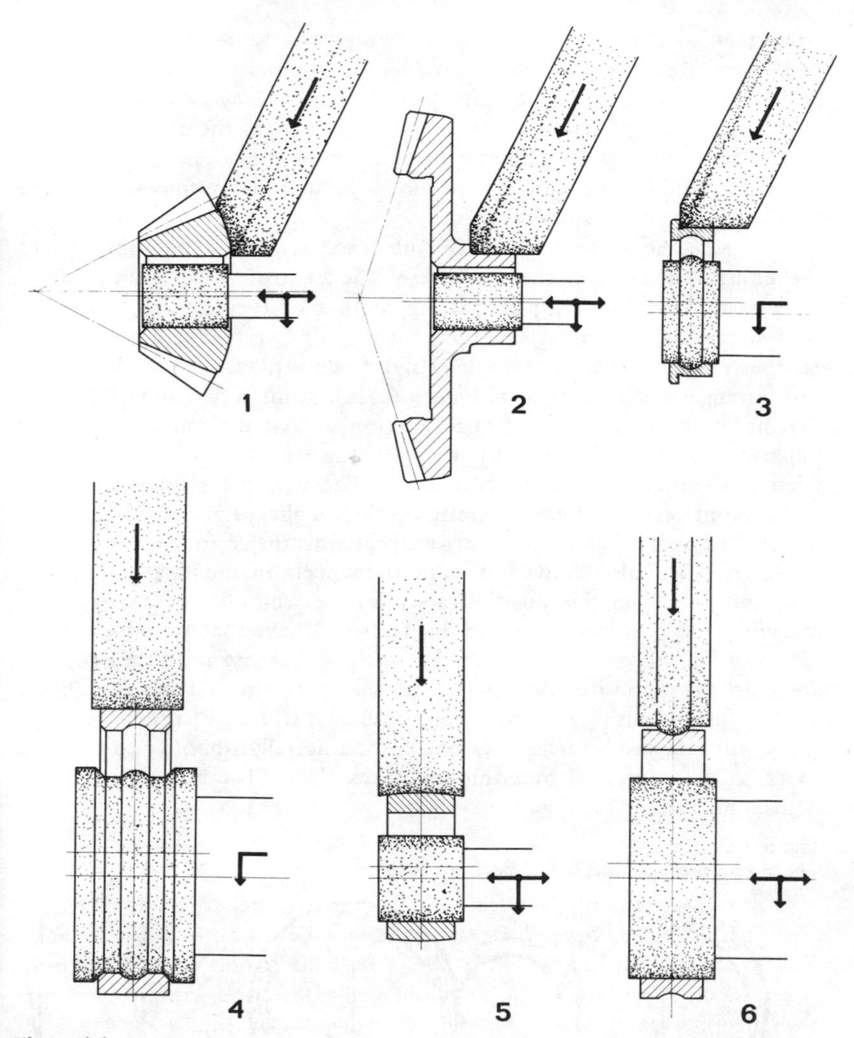

Figure 6.4.
Examples of high-production grinding
(By courtesy of C - G Grinding Machines, Paris)

The remaining examples (Figure 6.4, parts 4, 5 and 6) show simultaneous internal and external grinding, mainly with formed wheels, and were supplied by C-C Grinding Machines of Paris who have specialized in this field for many years.

CENTRELESS GRINDING

The centreless grinding machine takes a large part of the high-volume production of cylindrical parts. The process was formerly restricted to components of one or two plain diameters, but more recently work of quite intricate shapes has been produced at speeds difficult to reach by any other means. For example, plain cylindrical parts with only one diameter can be ground at a rate of up to 1½ miles of the diameter ground in an hour, whilst multi-diameter parts are ground at such speed that one centreless grinding machine can readily cope with the output of six multi-tool lathes or a multi-spindle lathe.

The machine consists primarily of three units: the grinding wheel, the control wheel and the work rest. The control wheel is made from an abrasive and bond approximating to that of the grinding wheel but runs at the same surface speed as the work, the latter being pressed simultaneously against the control wheel and downwards against the work rest. The arrangement is shown in Figure 6.5(a), from which it will be seen that both wheels run in the same direction so that the spinning tendancy imparted to the workpiece by the grinding wheel is controlled by the frictional contact on the control wheel. The workpiece takes the speed of the control wheel because static friction is always greater than sliding friction; indeed, it behaves as if it were geared to the control wheel.

Figure 6.5(a) also shows the work and wheels on one horizontal centre line, and, whilst in this position a workpiece which is presented to the machine with a perfectly round diameter will have that roundness maintained, a high spot on one side of the work will always produce a concave area directly opposite. A solution to this problem is shown in Figure 6.5(b), for, by raising the work centre relative to that of the wheel, high spots will still produce hollows but not diametrally opposite to the high spots, so that a gradual rounding out takes place. (The geometrical proof is given in Chapter 11, Figure 11:15.)

Figure 6.5.
The arrangement of wheels and work in centreless grinding

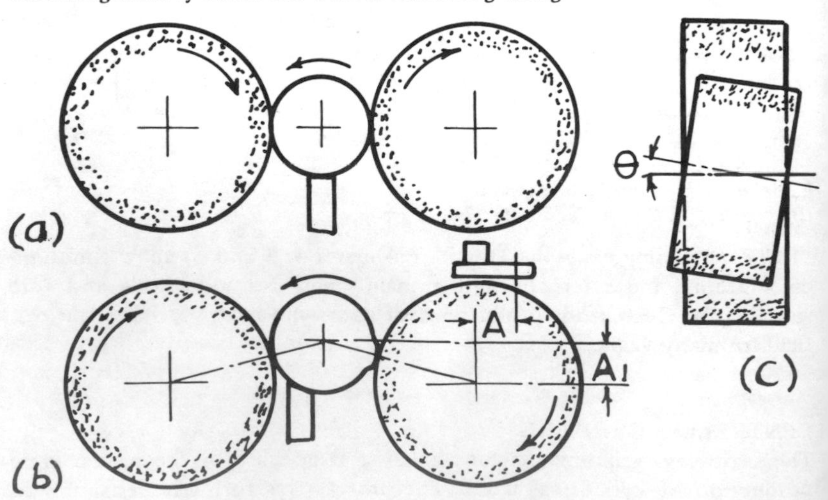

Centreless grinding may be divided into two general classes.

(1) Through-feed.

(2) In-feed.

In through-feed grinding the work is passed between the wheels, and no movement of the control wheel is needed except occasionally, to compensate for wheel wear. This method can clearly only be used for parallel work without shoulders although, of course, grooved work can be accepted where the grooves are not to be ground by the centreless machine. Axial movement of the workpiece is obtained by setting the axis of the control wheel at an angle as shown in Figure 6.5(c), and the rate of feed can be calculated from $ND\pi \sin\theta$, where D is the diameter of control wheel, N is the rotational frequency of the control wheel and θ is the angle of control wheel setting.

When the work is set above centre with the control wheel tilted, it is necessary to make a correction to the profile of the control wheel so as to ensure a full line contact between work and wheel over the full width of the wheel. This is accomplished by mounting the truing diamond in an adjustable slide which is off-set by the amount A, as shown in Figure 6.5(b), which is equal to A_1, the distance of the work above the wheel centres. The tilting of the wheel is taken care of by swivelling the slide carrying the diamond through an angle equal to the angle of tilt of the control wheel. Specially constructed centreless grinding machines can utilize a roller of the same finished diameter as the workpiece for dressing the wheels. The roller has fine industrial diamonds set into its periphery, and, as its geometry is precisely the same as that of the required workpiece, it can take the place of the workpiece and can dress the wheels to a perfect profile. Diamond roller dressing is dealt with later in this chapter.

The in-feed method is used for stepped work of lengths up to approximately the width of the wheels. The grinding action is that of plunge cutting, although the control wheel may be set to a very small angle to ensure that the work is kept in contact with a back stop. This is often coupled to an ejector which by lever or pneumatic means will eject the part from between the wheels.

PRODUCTION RATES

Just how high the production is from a centreless grinding machine can be seen from a few examples in Figure 6.6. A gudgeon pin (Figure 6.6(a)) is typical of the work produced on these machines, and, to finish such an item, several passes may have to be made. With a limit of ± 0.013 mm on the diameter an output of the order of 1,600 to 2,000 may be expected per day of 7½ h, but a reduction in the tolerance to 0.0025 may be expected to reduce the production rate by between 20 and 30%. On the other hand, with a wide tolerance of —0.025 mm, carbon rods for batteries (Figure 6.6(b)) are produced at the rate of more than 33,000

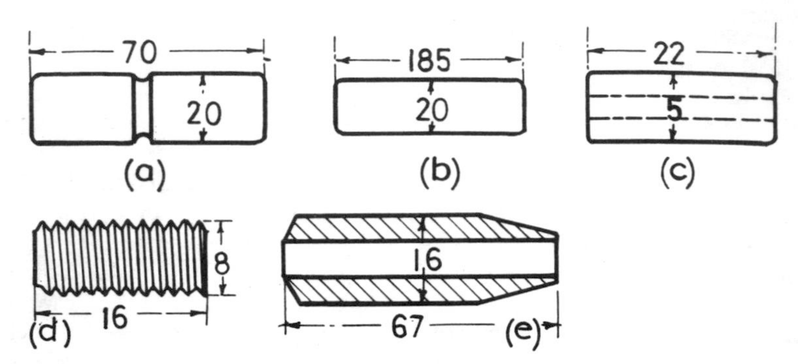

Figure 6.6.
Examples of work produced by centreless grinding

per hour. Ceramic parts such as Figure 6.6(c) are fed to the machine automatically from a rotary hopper and leave it at the rate of 3,000 per hour.

Such work as needle rollers constitutes a near-ideal job for centreless grinding, and in fact they are hard to finish by any other means except perhaps lap honing. A small roller with a stock removal of 0.117 mm and a limit of ±0.025 mm, with a normal finish, can be produced at the rate of 20,000 per hour. Powdered-iron cores for radio and television work (Figure 6.6(d)), manufactured from carbonyl iron powders and a suitable bond, are centreless ground to a tolerance of 0.025 mm at the rate of 25,000 per hour, the thread being centreless ground at a later operation. Cast iron valve guides to the size shown in Figure 6.6(e) are produced at the rate of 2,000 per hour with hopper feed, the stock removal being 0.18 mm and the limit on the diameter ±0.005 mm.

GRINDING OF SCREW THREADS

The process consists essentially of using a grinding wheel formed with one or more ribs around its periphery, these ribs having the form and pitch of the thread required. The grinding machine then acts in the same manner as a lathe used for screw cutting. These ribs are formed either by traversing the wheel with a shaped diamond dressing device or by crushing the form into the wheel by means of a hardened steel roller. The grinding wheel is usually set over to the helix angle of the thread to be formed although the wheel may be dressed to a compensated form to produce the thread with a vertical wheel.

One method of traverse dressing consists of a pantograph mechanism guiding a diamond as in Figure 6.7(a). It employs a hardened steel former A which is twenty-five times the size of the thread to be formed, a stylus twenty-five times the size of the diamond C and a ratio of 25:1 in the pantograph device. The diamond point is made to follow a path precisely the same shape as the former A but one-twenty-fifth of its size.

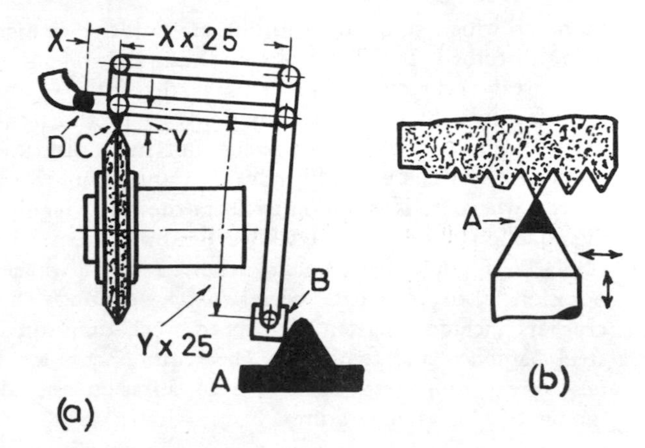

Figure 6.7.
Wheel-dressing methods for screw threads
(a) One method of traverse dressing:
A hardened steel former
C diamond
(b) Another method:
A diamond

Another method is shown in Figure 6.7(b). This arrangement is used for multi-rib wheels, the diamond being traversed across the face of the wheel by the lead screw of the machine, whilst interchangeable cams are used to feed the diamond into and out of the work in the correct manner to produce the thread form.

CRUSHING PROCESS

The crushing process (Figure 6.8(a)) which is a later development often replaces the diamond method on modern machines by the use of a hardened steel roller running on either centres or roller bearings. A roller with precision centres can be mounted in place of the workpiece and can thus offer a possible bonus of accuracy by dressing the wheel actually at the

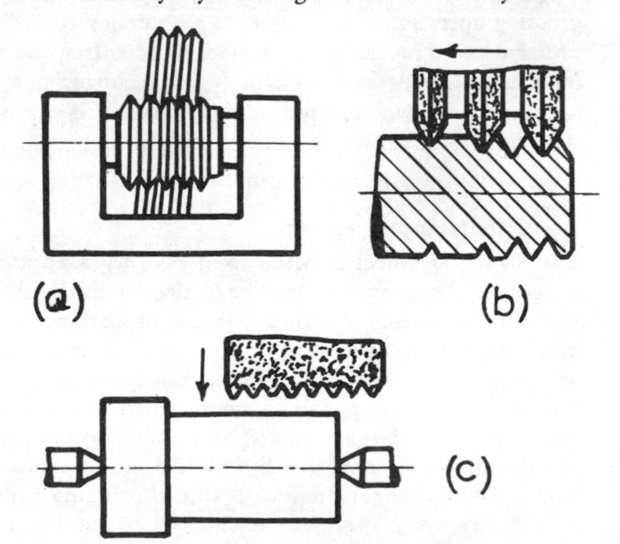

Figure 6.8.
The crushing process and grinding of screw threads

point at which it grinds without any of the elements of the machine being disturbed. It will be obvious that the roller's periphery is profiled with great accuracy to have ribs as the form of the thread. Different manufacturers have different methods of producing the relative rotation between grinding wheel and roller, but many use the roller to drive the grinding wheel as it is steadily advanced until a full thread form is produced on the wheel. It is common to form the ribs roughly with a single-point diamond on a new wheel whilst finishing them with the crusher, and similarly, as continued crushing tends to load a wheel, it is good practice occasionally to reform the wheel with a diamond. The materials used for crushers include tungsten high-speed steel, tungston carbide and carbon steel hardened and tempered. The requirements are for moderate hardness, great toughness, resistance to abrasion and ability to withstand moderately high temperatures.

GRINDING WHEELS FOR THREADS

Aluminium oxide wheels are usually used with a resin or vitrified bond. The first type will hold a fine edge longer than vitrified wheels owing to its inherent flexibility, but that flexibility may preclude its use on high-precision work, thus leaving the only choice as a vitrified bond. Wheels made from diamond powder with a rubber or plastics bond are often used for grinding threads in cemented carbides and other very hard materials.

The physical properties of the wheel vary according to the material and thread form, but it can be said that a wheel of fine texture with free-cutting properties is required, and for a general rule a grit size of 180 for coarse pitches, 220 for medium pitches and 280 for fine pitches may be employed. The correct cutting compound is important, as with many grinding operations, and oils with a tendency towards heaviness are to be preferred to thin oils which may tend to soften the wheel and may make its shape difficult to maintain. Soluble cutting oils are suitable for very accurate work where rapid heat dissipation is essential. The same oil is suitable for both grinding and crushing.

GRINDING OPERATIONS

The single-rib wheel is often used on very accurate workpieces such as gauges, for it generates less heat than a multi-rib wheel and therefore reduces the possibility of distortion or surface cracking. Equally, it is often more convenient for grinding up to a shoulder or narrow undercut, as a clearer view of the working area is possible. The wheel may be about 500 mm in diameter, and in operation it is inclined to be in alignment with the helix of the required thread. Good commerical ground threads of, say, 12TPI (2 mm) pitch and finer can be made at one pass of the wheel. One of the benefits of thread grinding springs from the ability of virtually every machine, at the end of the thread, is to withdraw the wheel from the work over a controlled arc of rotation, thus giving a strong thread run-out with a quality appearance.

If the run-out of the thread is not important and the pitch is comparatively coarse, the number of necessary passes may be reduced or the wheel life increased by using a three-ribbed wheel as shown in Figure 6.8(b). The first rib acts as a roughing rib, followed by an intermediate one which leaves, say, 0.15 mm on for a finishing rib which follows and perfects the profile. This type of wheel is useful for work such as engineers' taps as well as for long jobs which are unsuitable for the third method which is plunge grinding (Figure 6.8(c)).

In this method, the wheel is fed under a controlled feed into the rotating workpiece with the lead screw in operation; grinding continues for just over one revolution of the work after full depth of thread has been reached. Naturally, the length of the thread on the work must be less than the face width of the wheel, but the output rate from the machine is high.

Thread grinding eliminates the distortion which often occurs on heat-treated parts by permitting the threads to be machined after heat treatment, and it is a vital operation on such parts as machine tool spindles where threads and locknuts must be perfectly square to the axis of the spindle.

DIAMOND ROLLER DRESSING

As much as 25% of the ground parts made today are made on machines where the grinding wheel or wheels have been dressed by means of diamond-impregnated rollers. This is because the high-volume industries such as the motor car manufacturers require maximum output from their machines with the minimum of stoppages. If we consider the time taken for a single-point diamond dresser to traverse across a grinding wheel 100 mm wide, it will be clear that several components could be plunge ground in the same time. A diamond roller, which in effect grinds the grinding wheel, can reduce wheel dressing time from 3 to 4 min down to 6 or 8 secs. In the same way, the life of a diamond roller is infinitely longer than that of a single diamond, and therefore a machine can run for years, making the same part, without any stoppages as far as the dressing equipment is concerned; only wheels must be changed.

In the early days of diamond rollers, users had them made to precisely the dimensions of the finished component, and when dressing was needed they were placed in the grinding machine instead of the component. The wheel was plunged into the roller in the same way as when grinding, a small increment having been added to the dead stop to permit a small amount to be removed from the grinding wheel. To the C-C Grinding Machine company in Paris it soon became evident that there were several undesirable features of the system which mainly arose from the fact that a single diamond with its long dressing time exerted only small forces

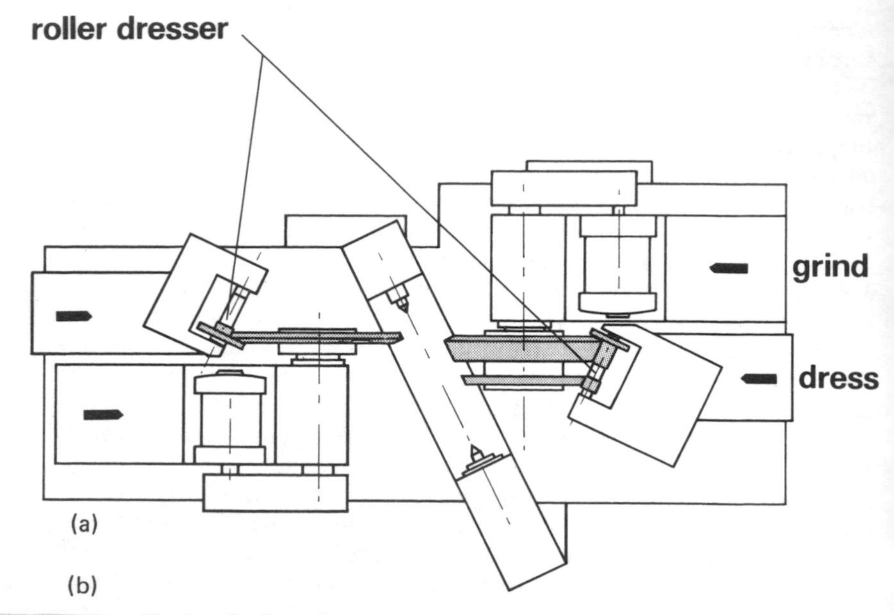
roller dresser
grind
dress
(a)
(b)

Figure 6.9.
(a) Diamond roller dressing of grinding wheel
(b) A diamond roller for dressing three wheels simultaneously
(By courtesy of C – C Grinding Machines, Henley-on-Thames)

on the machine, whilst the diamond roller with its quick dressing exerted large forces on the machine. Therefore, conventional grinding machines were not considered to be ideal for diamond roller dressing. That particular company which pioneered this philosophy of specially strong machines has since been joined by many others who also utilize diamond rollers, which are carried on a separate slide and are mounted on a machine specially designed for the workpiece. Figure 6.9(a) shows, diagrammatically, such a machine, and it will be seen that the roller still has the form of the component. Modern special-purpose machines often use several wheels on the same grinding head, and these can be dressed simultaneously with the same roller. Currently, C-C Grinding Machines and others are making machines of this calibre with multi-grinding heads, each with its wheel or wheels dressed in this manner. Figure 6.9(b) shows a diamond roller for dressing three grinding wheels simultaneously, the diamonds being set in a helical pattern with the roller diameters ranging from 6 to 16 in.

OTHER USES OF ABRASIVES

LAPPING

Lapping is a process used to finish cylindrical parts or flat ones to a high degree of accuracy and surface finish. Various abrasives are used including very fine grains of precious stones, but they are always much finer than those used for grinding wheels, and grits finer than 1200 are to be found in regular use. A typical application can be found in finishing a hydraulic plunger or piston after grinding (Figure 6.10). A cast iron tube, split several times longitudinally and capable of being compressed, is used and is referred to as the pot or lap. In the example shown the lap is tapered externally and can be forced into a taper in the main body, thus slightly contracting the bore; this permits the work which is reciprocated axially and is rotated to be polished to a precise size and perfect geometry. The abrasive, usually carred in a liquid, is introduced between the work and the lap either periodically or continually, and some users reverse the direction of rotation several times during the lapping process.

Flat workpieces are lapped either on a single flat plate or, if two parallel faces are required, between two lapping plates. The initial flatness and parallelism of the faces must be of a high order prior to lapping, and only final sizing and flattening with polishing is undertaken on the lapping machine. For flat parts, the laps are not always made from cast iron; a variety of plastics have been used, and at least one manufacturer of slip gauges has been known to use plates of pitch. Optical flats represent

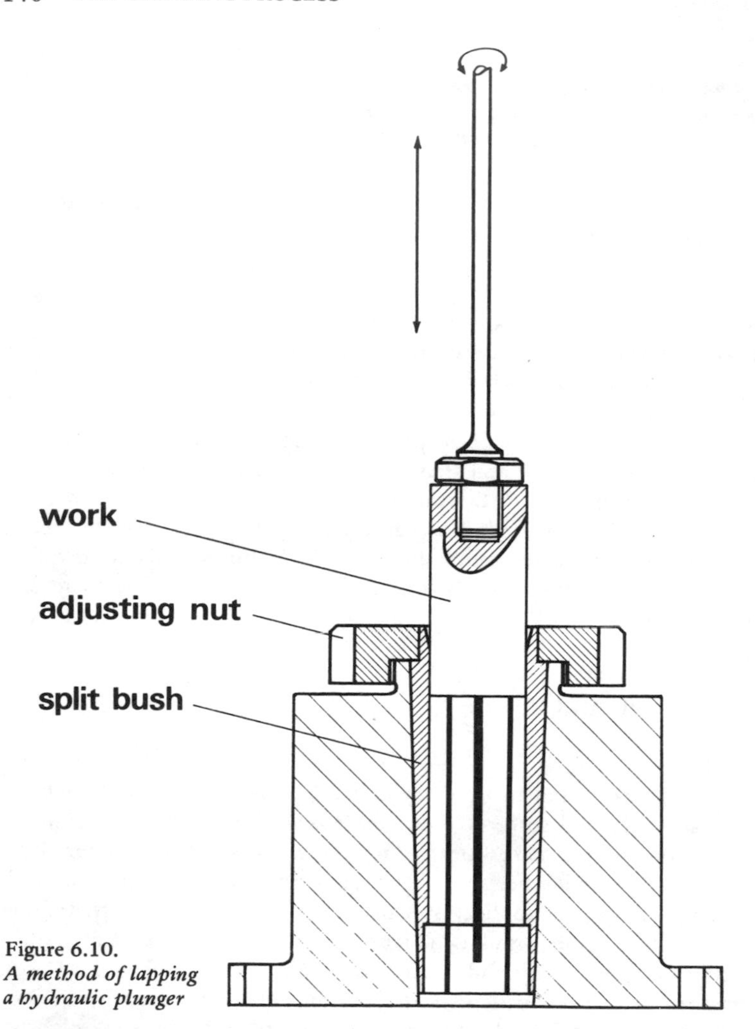

Figure 6.10.
A method of lapping a hydraulic plunger

the ultimate in quality work for this type of machine, and successively finer abrasives on successive machines must be used to obtain this immaculate standard.

LINISHING

External surfaces are sometimes finished by using a cloth belt which has been coated with an abrasive. Various grit grades are available, and the process is usually used where appearance is of more importance than dimensional accuracy although turbine blade manufacture has demanded this method of finishing for a number of years. Figure 6.11(a) shows that the belt is driven by a pulley and passes round a second free-wheeling pulley. Figure 6.11(b) shows a curved surface which is being linished

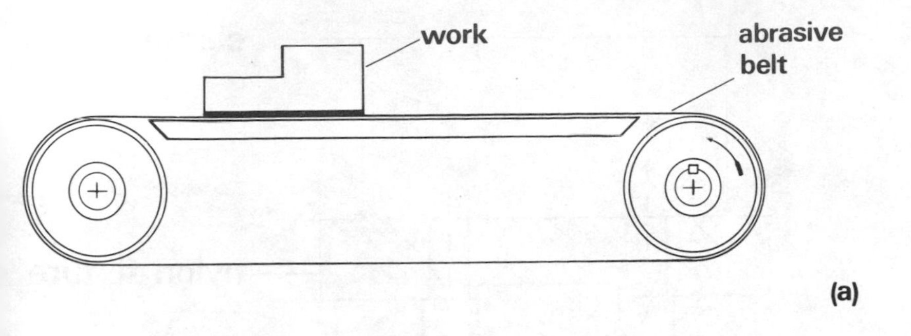

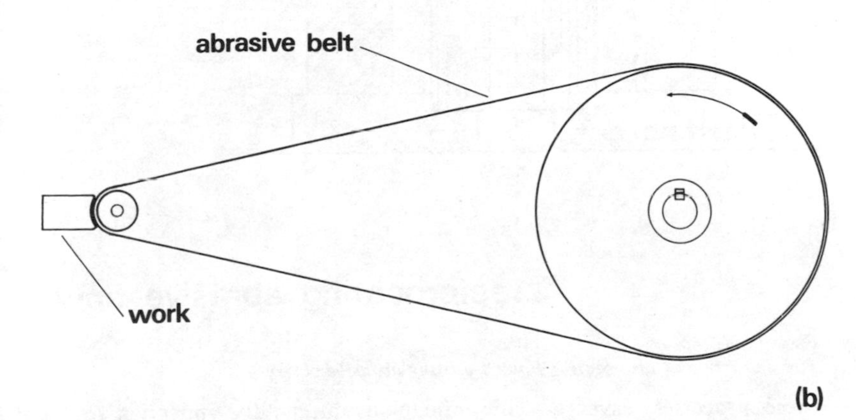

Figure 6.11.
The linishing operation which uses a cloth belt

by pressing the part gently against the belt, whilst in Figure 6.11(a) a flat surface is being processed.

SHOT AND VAPOUR BLASTING

Shot and vapour blasting are used for deburring of parts and for removing scale left after heat treatment or lengthy exposure. Many other finishing applications could be quoted, as the process leaves a fairly smooth finish all over the part with a matt grey appearance.

Basically the abrasive medium is propelled in a column of air at high speed to strike the workpiece; sharp corners are rapidly removed, and surface irregularities tend to be smoothed out; this is known as vapour blasting. Other media can be used for rough deburring or scale removal including small hard metal spheres which peen the corners of the work as well as wearing them away. Scale too can be removed in this way. Both

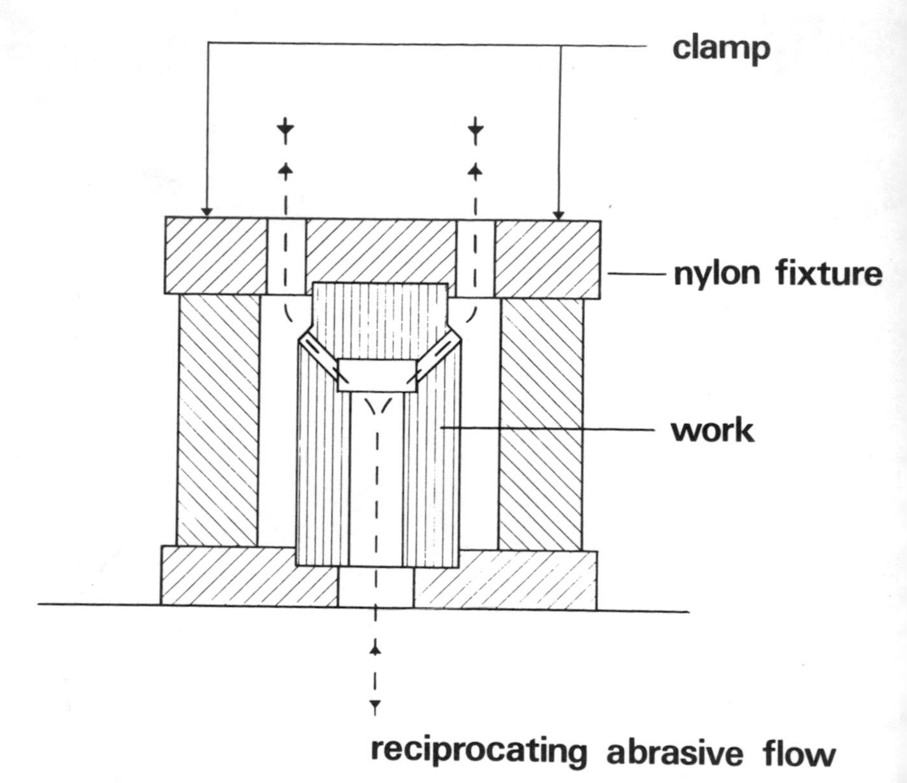

Figure 6.12.
The deburring of intersecting holes by using abrasive slurry

these processes leave the component absolutely dry and open to rapid rust attack, so that sometimes a final operation is made which shoots tiny pieces of lead at the part which gives a fine coating of the lead to it, thus providing temporary protection against rust.

BARRELLING AND VIBRATING

Barrelling and vibrating are abrasive machining processes (the former is sometimes known as tumbling) frequently used to finish large quantities of parts either by polishing them or by deburring them. For barrelling, the parts are loaded into a container together with the abrasive material, and the container is then sealed and rotated. With an action similar to that of a concrete mixer the parts and the abrasive are rubbed together; careful design of the container is needed to prevent the parts from being damaged by falling on one another, and the optimum load should always be used even if it is mainly abrasive. To avoid the damage mentioned above, the process of vibrating consists of loading the parts into an open-topped container together with the abrasive and of then vibrating the complete unit. Careful design of the container permits its charge to rotate slowly without any violent movement at all.

The abrasives used may vary between random-shaped chips of granite through man-made stone-like nuggets and conventional abrasives to fine polishing media with added pieces of lint to reduce accidental contact between workpieces. For deburring, the abrasive material shape can be very important as sharp-cornered material is needed to reach into intricate corners of the workpiece.

ABRASIVE DEBURRING OF HOLES

Frequently, in parts such as hydraulic valve blocks or diesel engine fuel systems, intersecting holes have to be drilled which show burrs to have been made where the second hole breaks into the first one. The more acute the angle of intersection the larger these burrs are likely to be, and, apart from the disturbance to oil or air flow caused by them, the possibility that a burr will become dislodged and will block a small orifice at some other part of the system cannot usually be tolerated. Deburring the hole intersections can be achieved by pumping an abrasive slurry backwards and forwards through the holes (Figure 6.12). A better abrasive medium is found in a mixture of grit and polymerized rubber which has the consistency of dough; it does not become packed up hard in odd corners of the work and is easily removed from the holes. This latter process was developed in the U.S.A. and is known as the Dynaflow process; suitable equipment for this work is now being made in the U.K.

7

COPYING, FORMING AND GENERATING SYSTEMS OF PRODUCTION

COPYING SYSTEMS

Profile turning of simple shapes can be carried out in a similar manner to that shown in Chapter 3 for taper turning. For cambered rollers, for example, the swivel bar is replaced by a slotted bar in which a roller moves to cause the saddle to move in an arcuate path. For the turning of cams the difficulty can be that the contour as it contacts the tool changes the top rake unless a rocking tool box is used. The Scrivener system obviates this difficulty by using a knife tool which travels axially along the lathe and machines the cam by a shaving cut (Figure 7.1). The principle is that of copy generating as shown in Figure 7.1(a), while Figure 7.1(b) shows how the tool is advanced and withdrawn from the work to give the correct contour. The master cam actuates the tool

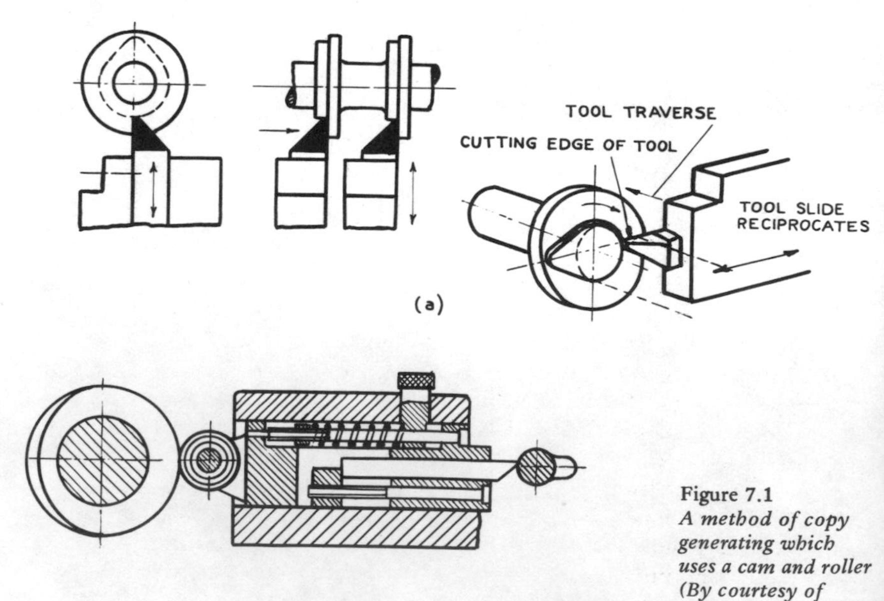

Figure 7.1
A method of copy generating which uses a cam and roller (By courtesy of Arthur Scrivener Ltd)

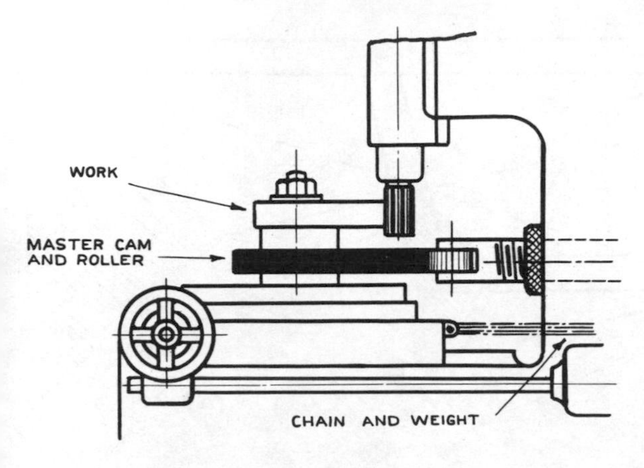

Figure 7.2.
A means of adapting a milling machine for cam production

holder through a ball bearing roller follower, the tool being kept in close contact by a spring. The radius of the tool point and the form traced by the point of the tool shall be that for which the master cam is designed.

A method of copying cams on a vertical milling machine is shown in Figure 7.2. A rotary table is driven from the feed shaft, and the table is free to move on the slides of the bed under the action of a chain and suspended weight. A screw passes through the column of the machine and carries a roller, against which the master cam makes contact. The master cam and work are set vertically on the table; this position allows the operator a full view of the machining. Cams with a constant rise can be milled by using a dividing head with change wheels to give the lead (see the discussion on universal milling in Volume 2 of this book).

METAL SPINNING

The simplest forming operations are those in which a shaped tool is fed straight into the work, as for forming shaped handles and bevelled surfaces, shown in Chapter 3. For spinning metal a former is required as in Figure 7.3 which is attached to a lathe faceplate. By using a suitable tool the metal is pressed onto the former and takes up the shape required. The metal shown in its original position at A is usually of copper, aluminium,

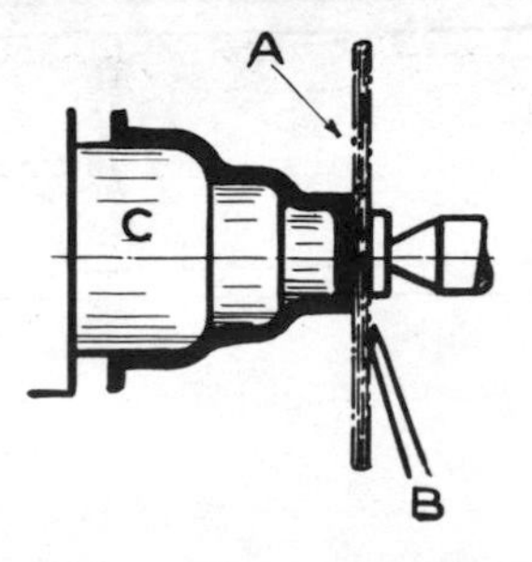

Figure 7.3.
The operation of metal spinning by using a former on a lathe faceplate:
A original position of metal
B rest
C former

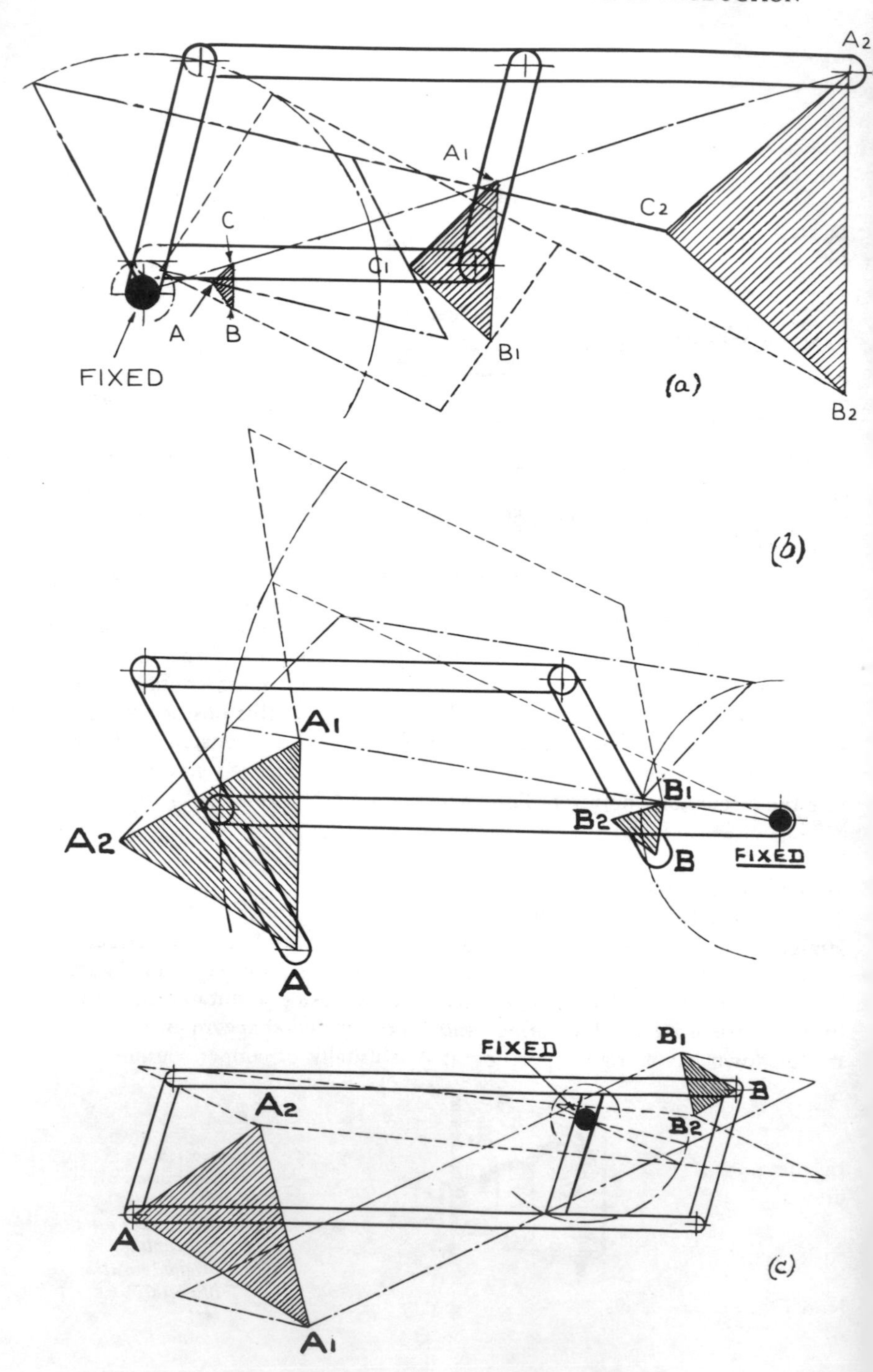
A2
A1
C
C1
C2
A
B
B1
B2
FIXED
(a)
(b)
A1
B1
A2
B2
B
A
FIXED
FIXED
B1
A2
B
B2
A
A1
(c)

Figure 7.4.
Types of linkages used on a pantograph mechanism

tin plate or mild steel. The tool which is held against a simple rest is shown at B. The former C may be of beech or steel which is used for large batches. The operation competes with power presses in many instances for the lathe and tool equipment is of the simplest and cheapest type. The articles produced include electric fire bowls, tea pots, trays, beakers and reflectors up to 6 ft (1.8m) in diameter.

PANTOGRAPH MECHANISM

A pantograph system is used for the production of dies and moulds which may entail intricate machining methods. It is based on the pantograph principle of similar figures in which corresponding sides or lines are equal or proportional; also a second feature of this principle is that their areas are proportional to the squares of their linear dimensions. In engineering practice the principle is extended to similar solids in which the matter of volume proportion arises. In this case the relation is that volumes of similar solids are proportional to the cubes of their linear dimensions. The best-known examples of pantograph mechanisms is that applied to engraving machines, and from the linkages shown in Figure 7.4(a), (b) and (c) it will be seen that, by rotating the frame around a fixed point while keeping the tracer point in contact with the large triangle, a cutter fixed in another member of the linkage will reproduce the same shape at a reduced scale in the work.

MATHEMATICAL TREATISE

From the parallelogram in Figure 7.5, to prove that points f, g and h must lie on the same straight line which passes through the fixed point e and that their motions will then be to their distances from the fixed point, move the point f to any other position f_1, when the linkage will be found to occupy the positions a_1, b_1, c_1 and d_1. Connect f_1 with e; then h_1, where f_1e crosses the link b_1c_1, can be proved to be the same distance from c_1, that h is from c, and the line hh_1 will be parallel to ff_1.

In the original position, since fd is parallel to hc,

$$\frac{fd}{hc} = \frac{de}{ce} = \frac{fe}{he}$$

In the second position, since f_1d_1 is parallel to h_1c_1 and since f_1e is drawn as a straight line, then

$$\frac{f_1d_1}{h_1c_1} = \frac{d_1e}{c_1e} = \frac{f_1e}{h_1e}$$

Now in these equations,

$$\frac{de}{ce} = \frac{d_1e}{c_1e}$$

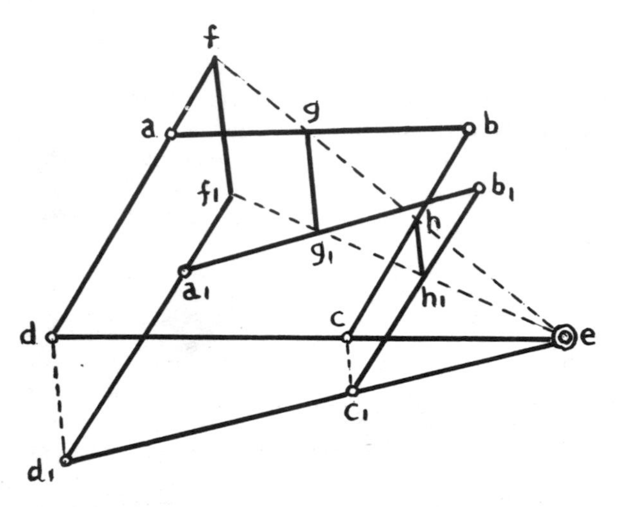

Figure 7.5.
A diagram showing the mathematical treatise of pantograph design

therefore

$$\frac{fd}{hc} = \frac{f_1d_1}{h_1c_1}$$

But fd $=$ f_1d_1 which gives hc $=$ h_1c_1 which proves that the point h has moved to h_1.

Also

$$\frac{fe}{he} = \frac{f_1e}{h_1e}$$

from which it follows that ff_1 is parallel to hh_1, and

$$\frac{ff_1}{hh_1} = \frac{fe}{he} = \frac{de}{ce}$$

or the motions are proportional to the distances of the points f and h from point e.

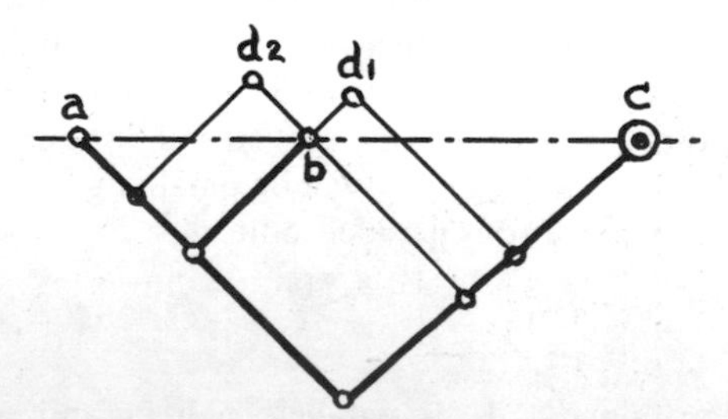

Figure 7.6.
A method of constructing pantograph linkages

To connect two points such as a and b in Figure 7.6 by a pantograph so that their motions shall be parallel and similar and in a given ratio, then firstly the fixed points must be on the continuation of the straight line ab and must be so located that the ratio of ac to ab is the same as the desired ratio of the motion of a to b. After locating c, an infinite number of pantographs can be drawn, but care must be taken that the links are so proportioned as to allow the desired magnitude and direction of motion. It is interesting to note that, if b were the fixed point, a and c would move in opposite directions. It can be shown as before that their motions would be parallel and in the same ratio as ab to bc.

The pantograph machine of to-day has developed into a high-speed milling machine. Figure 7.7 shows how a tilting pantograph system

Figure 7.7.
A pantograph system with a rocking movement for deep copying:
A arms C link system E second tracer point G holder J work
B pivot D tool carrier F slide H position K holder

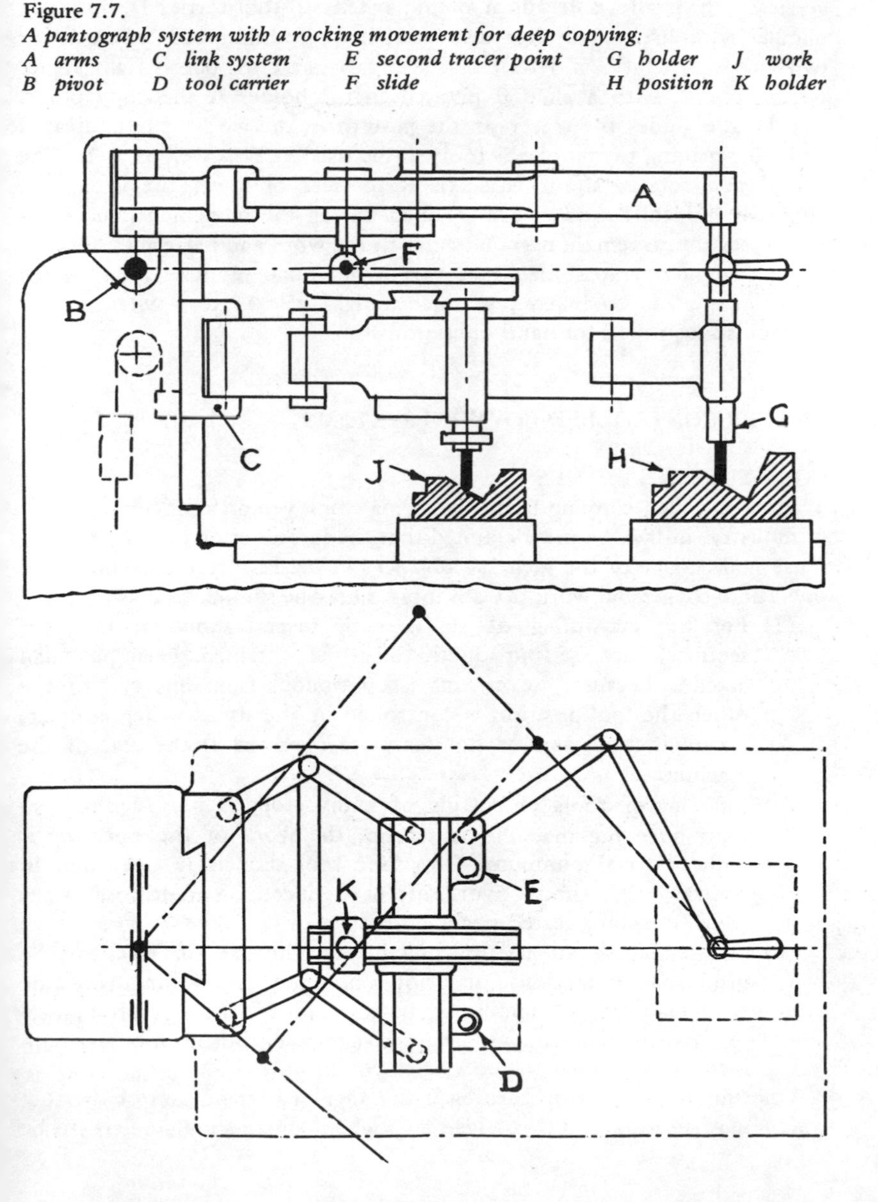

mounted for a rocking movement about an axis parallel to the plane of the system can be used so that the relative position of the tracer and the tool in respect to the work is not changed. To effect this, they are connected at the tracer and copying points by a pivot in such a manner that movement of the pantograph system about the axis ensures that the tool and tracer always move perpendicularly to the work system.

The arms A pivot about B, while the link system C can be moved vertically. This allows freedom of movement to tool carrier D, but only parallel with itself. This bar also contains a second tracer point E for working with a ratio of 1 to 1 and also carries a cross bar at right angles to the pin B, with a slide F pivoted to the holder at the copy point. The handle guides the tracer on the pattern H, the work being indicated at J. For making two copies a tool can be used at E as well as at D. The pantograph rotates about the axis B to raise or lower the tool, but, although holders K and G are inclined during the movement, the tools and tracer always remain perpendicular to the work and pattern.

Pantograph-operated machines vary from table models which weigh as little as 30 lb to machines which weigh over 15 tons where power control is substituted for hand operations.

TRACER-CONTROLLED COPYING SYSTEMS

COPY TURNING LATHES

Tracer-controlled copying lathes are being employed on an increasing scale in industry; in fact it can be claimed that the introduction of copy turning must rank as one of the greatest advances in the history of machine tool operation. On general work the advantages may be summarized as follows.

(1) For the machining of shafts with several shoulders or taper sections, increased production can be obtained, even on small batches, because the cutting is continuous from one end to the other; the tool position is controlled by the stylus which contacts a cylindrical master or flat template mounted at the rear of the machine.

(2) Indexing of tools or setting of many stops is not required, so that, once the machine is set up, the work of the operator is reduced to a minimum. Only one tool slide is in operation in contrast with similar work often produced on multi-tool lathes which use complicated tooling arrangements.

(3) Square faces can be machined, or spherical surfaces can be produced; operations are not restricted to turning, for the advantages apply equally to boring and even to the production of non-circular shapes, as, for example, bottle moulds with intricate engraving.

One limitation to copy turning is the angular presentation of the tool which may introduce difficulties, say, when square shoulders are to be

Figure 7.8.
A lathe with a hydraulic tracer-controlled copying system
(By courtesy of T.S. Harrison & Son Ltd, Heckmondwike)

turned on a shaft with decreasing diameters. These difficulties are easily overcome by a second setting of the workpiece or by making use of the front tool post for certain operations.

Copying systems may be hydraulic, electric or electronically controlled, but Figure 7.8 shows a lathe by T.S. Harrison & Sons Ltd with hydraulic equipment. This comprises the oil tank and motor-driven pump unit at the end of the bed, with connecting pipes to the tracer control valve at the rear of the saddle. A cylindrical template is shown mounted between centres at the rear of the bed, which is a duplicate of the workpiece shown mounted on the lathe. Turning is from the rear tool post when copying, thus leaving the front tool post available for other operations to supplement the copy turning.

Another feature of the Harrison lathe is that a reversible motor is fitted so that the copy tool does not need to be inverted in the tool rest and cutting forces are directed downwards on all the slides, thus damping out any tendency to produce vibrations.

ANGULAR TOOL PRESENTATION

Figure 7.9(a) shows that, with the copy slide set at an angle of 30° to the vertical and with the traverse operating in the direction indicated by the arrow, shoulders up to 90° can be produced, but falling shoulders are limited to 30°. The relationship of the two movements is

$$\frac{\text{movement of ram}}{\text{movement of saddle}} = \frac{2}{1}$$

thus, if the ram retracts twice as fast as the saddle traverse, a square shoulder will be produced.

The effect of the angle of entry can be seen as follows. Let V_l be the speed of the longitudinal feed, V_t the speed of the transverse feed and V_c the speed of the cutting tool slide; then if $\alpha = 30°$,

$$V_t = \frac{V_l}{\sin\alpha} = \frac{V_l}{0.5} = 2V_l$$

and

$$V_t = \frac{V_l}{\tan\alpha} = \frac{V_l}{0.577} = 1.73\,V_l$$

and

$$V_c = \frac{V_l}{\sin\alpha} = \frac{V_l}{0.5} = 2\,V_l$$

Figure 7.9 shows that the ratio of the saddle feed when the tool is moving up the workpiece is given by

$$\frac{W_f}{S_f} = \frac{\cos\beta}{\cos(\alpha - \beta)}$$

Figure 7.9.
A diagram illustrating the angular tool presentation when copying

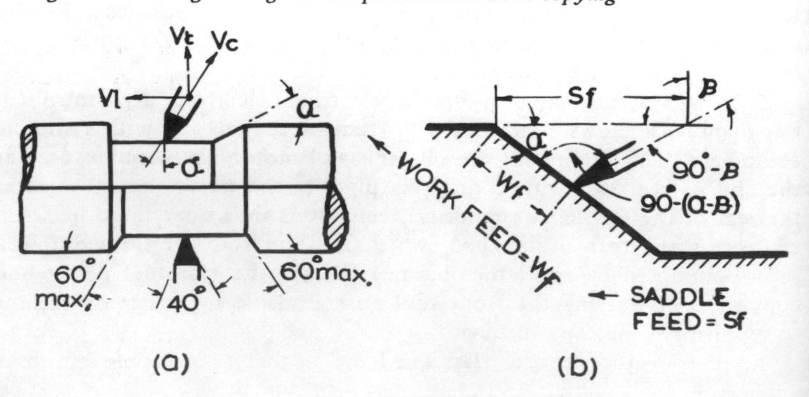

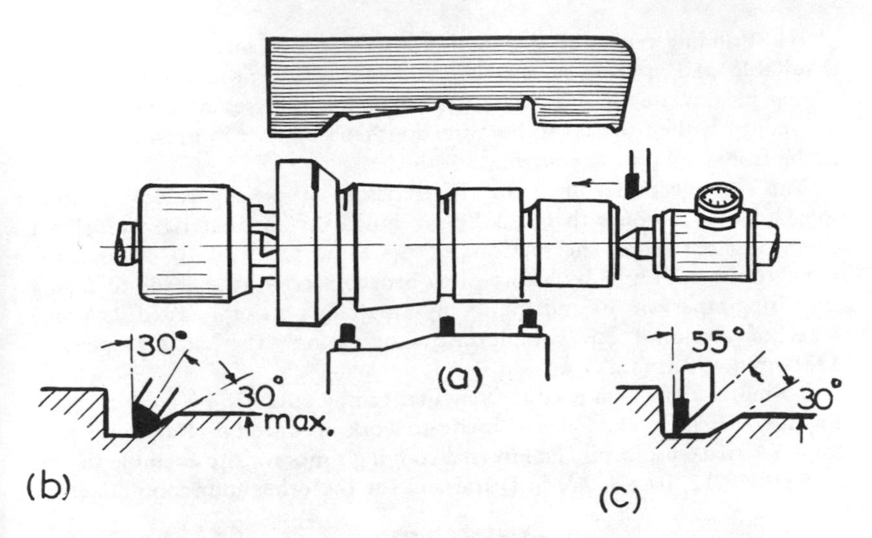

Figure 7.10.
An example of copy turning by using a sheet metal template

where β is the angular setting of the tool from a line perpendicular to the axis of the workpiece, α the angle to bc produced on the workpiece, S_f the saddle feed of the lathe and W_f the workpiece feed.

$$\frac{W_f}{\sin(90-\beta)} = \frac{S_f}{\sin[90-(\alpha-\beta)]}$$

Therefore

$$\frac{W_f}{S_f} = \frac{\sin(90-\beta)}{\sin[90-(\alpha-\beta)]} = \frac{\cos\beta}{\cos(\alpha-\beta)}$$

A typical set-up for copy turning is shown in Figure 7.10(a). It shows the advantage of using a Kosta driver so that the component can be turned over its full length, by using a pressure gauge on the tailstock centre. The copying tool mounted in the rear tool post travels from the position shown over the full contour, as the tool is controlled from the stylus contacting the sheet metal template.

Figure 7.11.
A test piece showing contours produced on a Harrison lathe

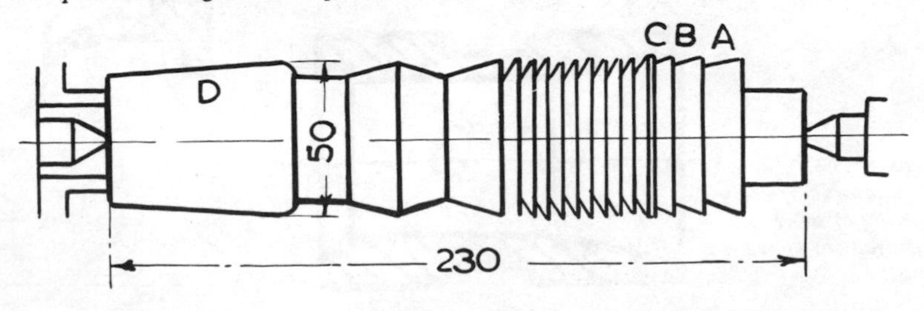

If a grinding recess of the section indicated in Figure 7.10(b) and (c) is suitable, and in most cases it is, the recess can be produced by the tool during its traverse. If, however, square-section recesses or other grooves are required, then these can be turned either singly or in unison by tools in the front tool post as shown.

The test piece machined on the Harrison lathes (Figure 7.11) shows some of the contours that can be machined on bar material by using a cylindrical template. The sections of A, B and C are of 10, 15 and 20° respectively, followed by a fine-pitch broach section; this leads to falling and rising tapers of 30° and thence by a parallel portion to a No. 1 Morse taper of section D to complete the operation. The cutting speed is 333ft/min (90m/min).

To indicate the high productivity that can be obtained by copy turning, Figure 7.12 shows examples of intricate work produced on Harrison lathes for USA space projects. The internal copying times for the example shown in Figure 7.12, part 1, are in Dural and for the other four components in

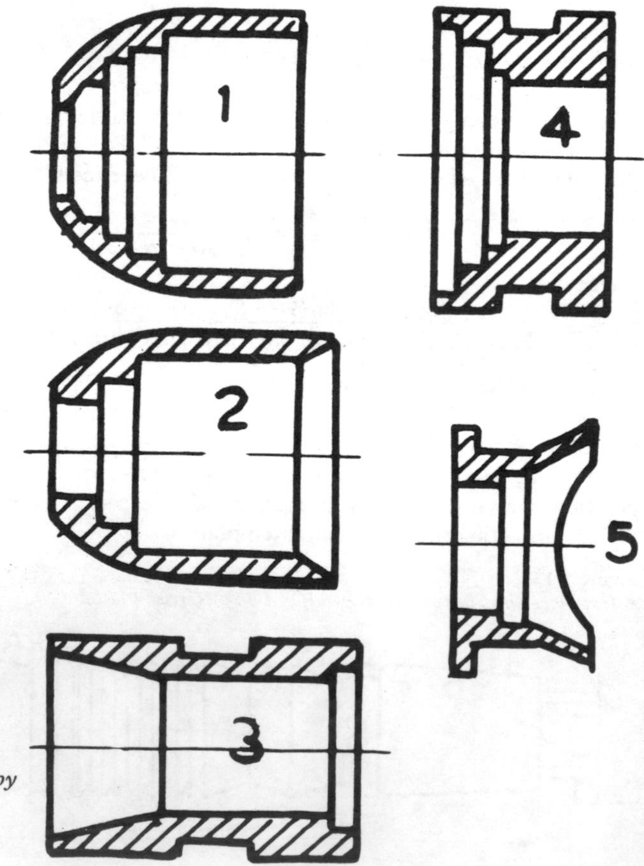

Figure 7.12.
An example of copy turning of work required for space projects

steel. Roughing and finishing cuts were used at a feed rate of 0.02 in/rev (0.5 mm/rev). The production times are as follows: 1 min 10 s for the example shown in Figure 7.12, part 1; 1 min 30 s for that shown in part 2; 1 min 30s for that shown in part 3; 2 min for that shown in part 4; 1 min 10 s for that shown in part 5. The external times were about 1½ times as long as for the internal machining. The production figures are, for part 1, 400 components per day instead of 100; for parts 2, 3 and 4, 200 components per day instead of a previous 80; for part 5, 400 components per day instead of a previous 135.

COPY TURNING BY USING HYDRAULIC CONTROL

A copy turning system which uses hydraulic control is shown on the Harrison lathe, in Figure 7.13 and is detailed to show the hydraulic valve control. The stylus contacts the template mounted on a bracket at the rear of the lathe bed. When the stylus is disengaged from the template, a spring forces the piston valve in the direction of the template so that the inlet to the ram chamber and the exhaust line from it are opened.

If oil is now supplied to the piston valve of the tracer, the tool slide will move towards the template and the workpiece. When the stylus contacts the template, the valve is forced back against the spring and closes the line to the chambers of the ram. If the longitudinal traverse is now engaged, on the assumption that the stylus is in contact with a

Figure 7.13.
The details of valve control as used on a Harrison copying lathe

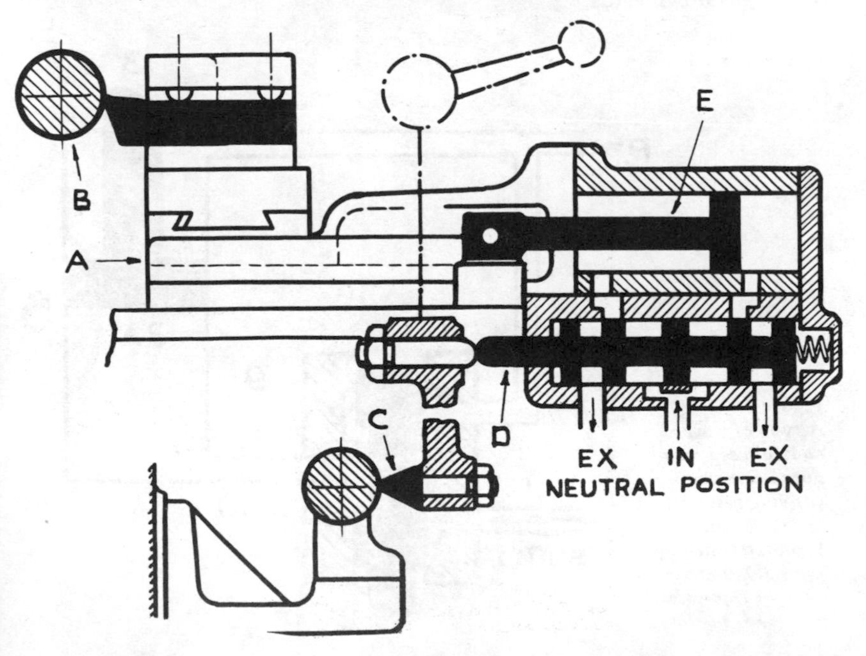

parallel shaft, or a straight section of the template, the tool will follow a straight path along the lathe. If now the stylus encounters an increasing taper section, the valve will open in a reverse direction, and the tracer and tool will move away from the template. On completion of the taper and at the return to a parallel section, the spring forces the stylus forward again so that the control ports are covered as before. It will be apparent that successful operation depends on accurate matching of the valve ports.

POTENTIOMETER SYSTEM

The potentiometer system is an alternative design (Figure 7.14) in which the valve acts as a hydraulic potentiometer to vary the pressure in pipe 8. In the position shown the openings 9 and 10 are equal, and in consequence the continuous flow of oil through the valve and back to the sump through pipe 11 causes half the supply pressure P to drop across opening 9 and half across opening 10. The piston 13 is so arranged that, if the effective area on the top side is A, the area on the bottom side is $2A$. The pressure on the top side is therefore $P \times A = PA$ and on the bottom side $(P/2) \times 2A = PA$, i.e. the forces are equal. In the position shown the piston is therefore at rest, but, if the pilot valve member 12 is displaced upwards, the opening 9 is increased and the opening 10 is reduced by equal amounts. Consequently the pressure in pipe 8 increases, and piston 13 moves upwards. Conversely, a downward displacement of member 12 decreases the pressure in pipe 8, and piston 13 moves downwards.

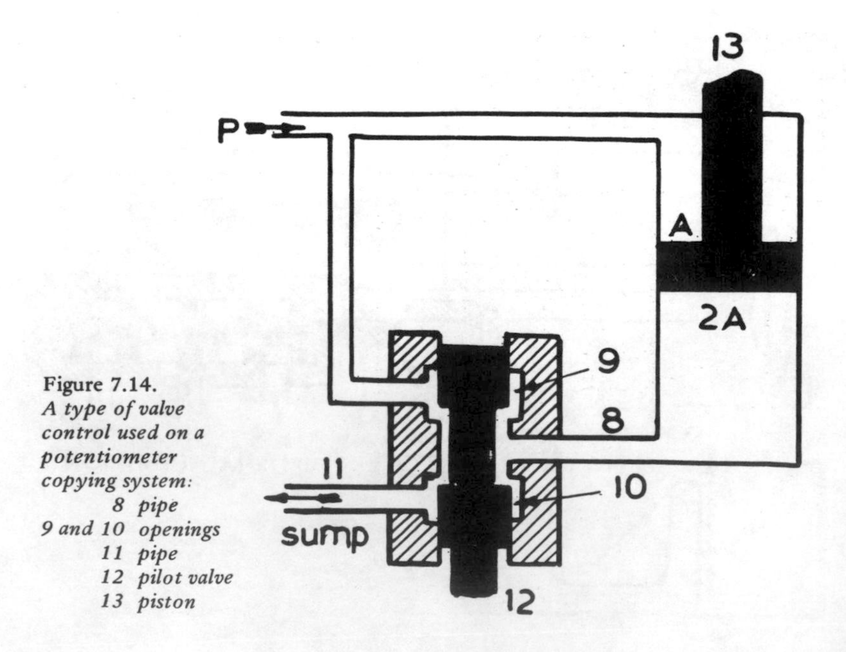

Figure 7.14.
A type of valve control used on a potentiometer copying system:
8 pipe
9 and 10 openings
11 pipe
12 pilot valve
13 piston

Variations in oil consumption are small, and it is easy to obtain a constant-pressure source. A disadvantage is that the effective piston area is not $2A$ but A; consequently, for a given maximum potential force, the diameter of the cylinder is greater than the conventional type by a factor of 1.4 for the same pressure P.

MILLING OPERATIONS

The procedure using a tracer finger is to place the master and mould blank on the machine table in line beneath the tracer and cutter. Figure 7.15 shows a Cincinnati design which controls the up and down movements of the vertical slide carrying the tracer and cutter and which also controls the rate but not the direction of the horizontal movement of the table carrying the work and master. The axis of the tracer stylus is in the plane of the contour, thus making the arrangement suitable for such operations as die sinking. Sideways pressure on the stylus rocks it on its spherical seating, and this is converted to axial movement of the piston valve by the ball seating arrangement between the stylus and valve.

Movement of the valve directs the flow of oil to the vertical cylinder and controls the rise and fall of the slide so that the cutter follows the path of the stylus. Tracer deflection also determines the reduction of the horizontal movement of the table, master and work by means of the throttling ports in the tracer valve connected to the hydraulic circuit of the horizontal cylinder. When a sharp slope on the master is

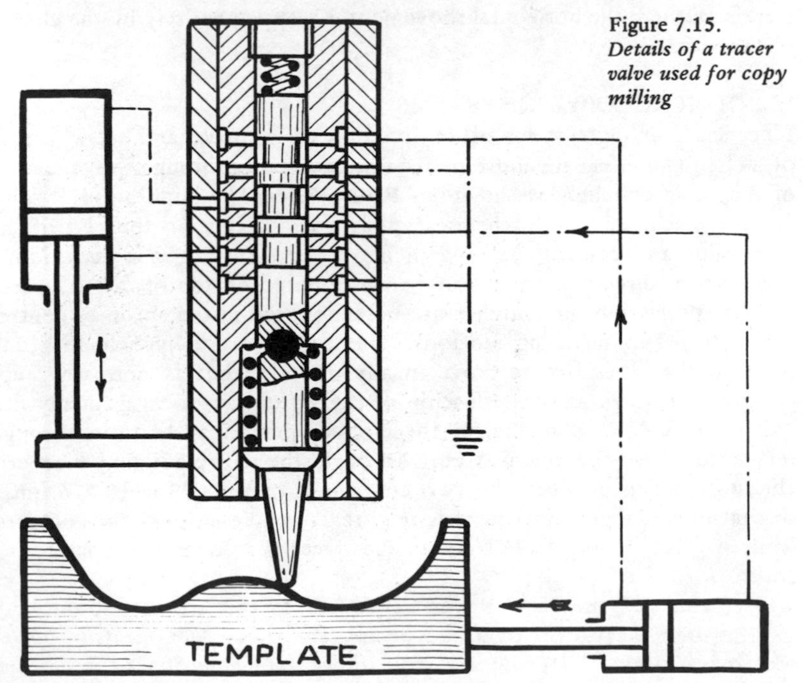

Figure 7.15. *Details of a tracer valve used for copy milling*

Figure 7.16.
An electrical copying system which uses magnetic clutch control

encountered, the horizontal movement is decreased, and, when a vertical step is reached, the horizontal movement is stopped entirely by the closing of the ports.

ELECTRICAL COPYING SYSTEMS

Electrical copying systems often function on a 'make and break' action of a circuit to effect the movement of a cutting tool through the operation of magnetic clutches, as shown in Figure 7.16. The application is for a planing machine; the clutches are mounted in a box on the end of the cross slide and control the movement of the tool box in a longitudinal and vertical direction, the stylus being mounted on top of the slide. For a lathe application the clutches are best mounted in the apron to control the sliding and surfacing motions. The tracer is usually held in a ball joint so that it is free to move in any direction but is normally kept central by springs. Lever deflection of the stylus causes engagement with either 'in' or 'out' contact and the circuit operates at 14 Volts through relays to close the main circuit through the corresponding magnetic clutch. The gap between the two contacts is only 0.003 in (0.076 mm), so that only a slight movement is required to make or break the contacts, while a force of 4 ozf (113 gf) on the tracer is sufficient to operate the controls.

A feature of the Heid tracer control is that it works without any intermediate relays or remote control switches. The electric power required is about 50W d.c., i.e. the power consumption of a normal lamp bulb. To translate the deflection of the stylus into feed movements, electrical impulses of varying duration are transmitted to the clutches. The tracer contacts engage and disengage the clutches, while the feed motor runs continuously. The tracer unit incorporates three sets of contacts. An adjusting screw on the tracer is used to vary the gap in the contacts to suit individual requirements; the closer the gap, the more accurately is the contour reproduced. The low-voltage supply makes possible a contact gap of less than 0.004 in (0.102 mm) in the tracer.

ELECTRICAL LIMITATIONS

In the normal electrical system the operating time of relays, contactors and magnetic clutches may reduce the rapidity of the action of contouring. Figure 7.17(a) shows how the tracer moves over the template in steps, and with it the tool mounted on the same slide. Deviations from the true profile are exaggerated to demonstrate the effect, for in practice the deviations are minute and can be reduced by an increase in the switching time, by a reduction in the feed rate or by altering the ratio between the longitudinal and transverse feed rates. The response of an electric tracer control is a function of the tool displacement per unit of time. If we assume the machining of the contour as in Figure 7.17(b) with a constant feed rate V_l in the longitudinal direction, then various feed rates in the transverse direction will be necessary to provide the various slopes. Next we assume that the two feeds are so related that the 75° slope is traversed without a correction impulse from the tracer. If the time taken by an impulse t_k is 0.1 s, the feed is 4¾ in/min and the 75° slope is explored steplessly, the transverse feed rate will be

$$V_c = \frac{V_l}{\tan 15^\circ} = \frac{4.75}{60 \times 0.2679} = 0.294 \text{ in/s}$$

Figure 7.17.
The limitation of electrical devices for tracer copying

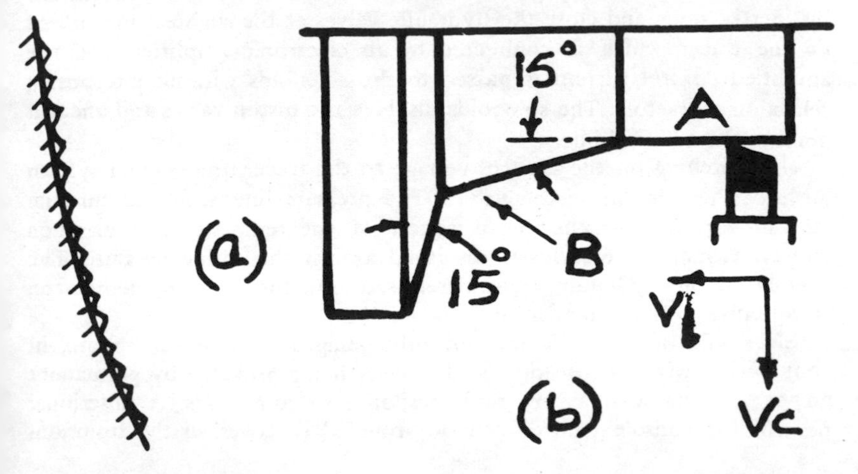

If the tracer switches at any point on the parallel part A, then the error of tolerance T, if we allow a switching delay of $t_k = 0.1$ s, is expressed by taking into account the time elapsing before the machine applies a correction, as $T_{75°} = 0.0294$ in.

On slope B the tolerance is

$$T_{15°} = \frac{V_l}{\tan 15°} - V_l \tan 15° = \frac{0.0078}{0.2679} - 0.0078 \times 0.2679 = 0.029 \text{ in}$$

since in a time of 0.1 s the tool travels 0.0078 in longitudinally.

Modern developments include means for reducing the control currents so that pitting of the contacts is prevented for long periods of use. This enables an increase in the switching frequency to be made and also enables a good finish to be obtained on the work. The location of the clutches should be as near as possible to the tool, so reducing the accelerating time of the rotating masses of gears and shafts to a minimum.

ELECTRONIC OPERATION

Tracer-controlled copying is applicable to large machines and heavy workpieces. Figure 7.18 illustrates this; the operation is that of machining deep grooves at both ends of a heavy roller, the sheet metal template and the electronic head being mounted at the rear of the lathe. With one system the stylus is capable of deflection which is resolved electrically into two components which are parallel to the sliding and surfacing motions. The resulting electrical signals are amplified electronically and are arranged to control both the directions of rotation and speeds of two feed motors in such a way that the stylus moves along the contour of the template and the tool reproduces a similar contour.

The system shown uses electro-hydraulic control in which the tracer transmits minute electrical impulses to an electronic control unit which operates small magnetically operated hydraulic valves, avoiding any mechanical contact. The high-vacuum valves of the control unit, unaffected by inertia, open and close the hydraulic valves at the slightest movement of the tracer which is connected to an electronic amplifier, and the amplified control current is passed to the solenoids without interposing relays or contactors. The solenoids displace the piston valves and uncover ports to the main cylinder.

On switching on the control voltage to the tracer, the magnet system pushes in one of the piston valves. The pressure line is opened, and the ram moves the slide in the direction of the template; finally, when contact is made, the stylus is displaced against the spring pressure. The magnet in the hydraulic valve is released, and the spring-loaded piston valve returns to its zero position.

Electronic tracer heads are currently being used, more and more, in conjunction with electric drives, the feeds being provided by permanent magnet motors with a very rapid response to commands. A functional pendant or console control can be provided to cover all the functions

Figure 7.18.
An electronic tracer copying system used on large lathe

of the machine without embodying hydraulic pipework to the control station.

Figure 7.19 illustrates such a control station from an Oerlikon die-sinking machine; the lower panel includes all normal machine controls, feed direction, rate and spindle controls together with a coolant switch. The upper panel is the control board for the tracer unit, and it can be

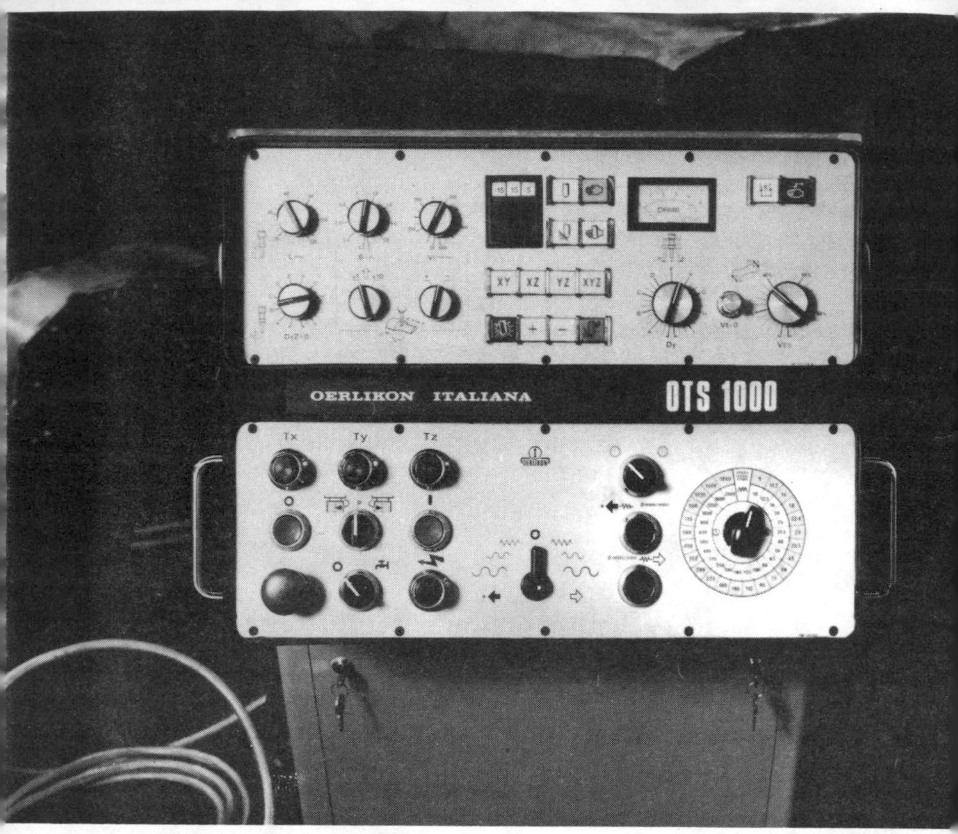

Figure 7.19.
A control station for a die-sinking machine

seen that any two or all three axes can be selected to be in operation at one time, thus permitting contouring to take place on up to four sides of a workpiece at one setting. Controls are also provided to pre-set the amount of stylus deflection being used during roughing operations and a fine one for finishing. In order to make the machine as economical as possible, the depth of cut and the increment at each pass is also pre-set on this board, thus defining the chip section with accuracy during roughing; for finishing the values would be reduced to obtain the required accuracy and surface finish on the workpiece.

Figure 7.20 shows the Oerlikon OTS 1000 tracer head with a stylus mounted in it. Deflection can take place either axially or radially in any plane; inductive elements within the head sense deflection and send signals to the thyristor controls for the three feed motors.

If the feed drives are sufficiently powerful, very heavy roughing operations can be undertaken with such a system, or, alternatively, several milling heads can be used simultaneously. Figure 7.21 shows the Oerlikon OTS 1000 system controlling a machine with four heads, thus producing four workpieces simultaneously; the workpieces shown are partially roughed masters for ski boot soles.

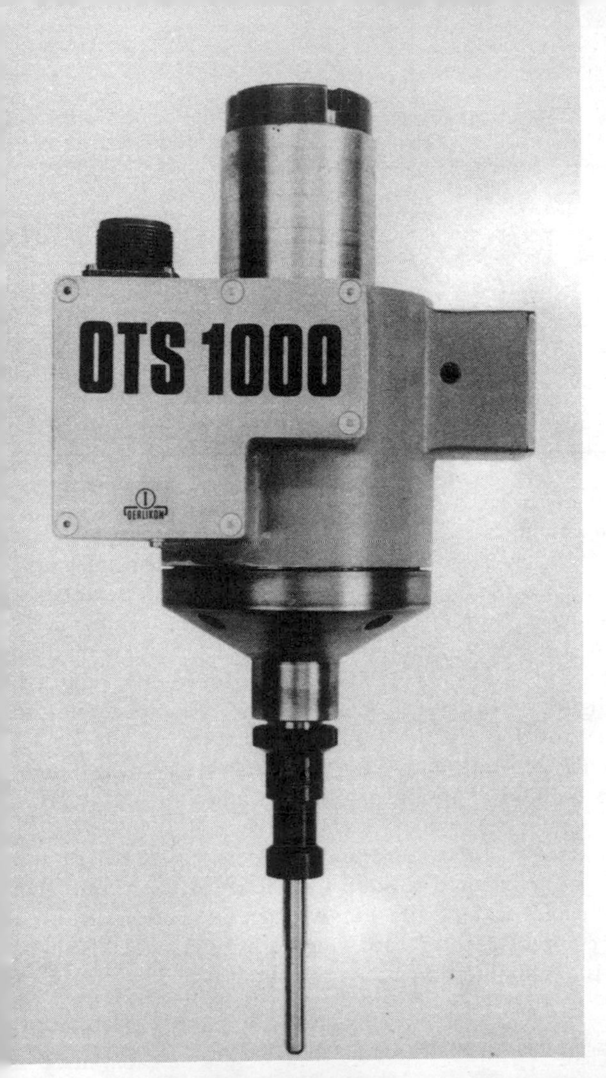

Figure 7.20
An Oerlikon tracer head and stylus

Figure 7.21.
A die-sinking machine with four heads for producing dies for ski boot soles

8

TOOL-HOLDING EQUIPMENT AND TIME STUDY

The high productive capabilities of a machine tool may be impaired if the tool-holding equipment is restricting the machining operations. Thus consideration should be given to what is available for efficient metal removal by using suitable tooling for a given sequence of operations. The method of holding a tool on a lathe, for example, was formerly by a double strap, or by the American-type tool post; the first method is clumsy, and the second lacks rigidity. Where a single tool only is required, then the design of Figure 8.1(a) gives a good support, is compact and can be set at an angular position by the segment gear teeth on the holder and a taper bolt which prevents movement under heavy cutting. There is also a bored hole to hold a boring bar; this in addition to the lathe turning tool.

Two types of holders for form tools are shown in Figure 8.1(b) and Figure 8.1(c); the first one is for a dovetail tool designed to operate on a rear tool post. The tool has long life but is expensive and difficult to produce, so that the circular form tool (Figure 8.1(c)), which can be shaped in a lathe, is preferable and also has a long life. The centre of

Figure 8.1.
Types of tool holders used for lathe turning and forming:
S serrations

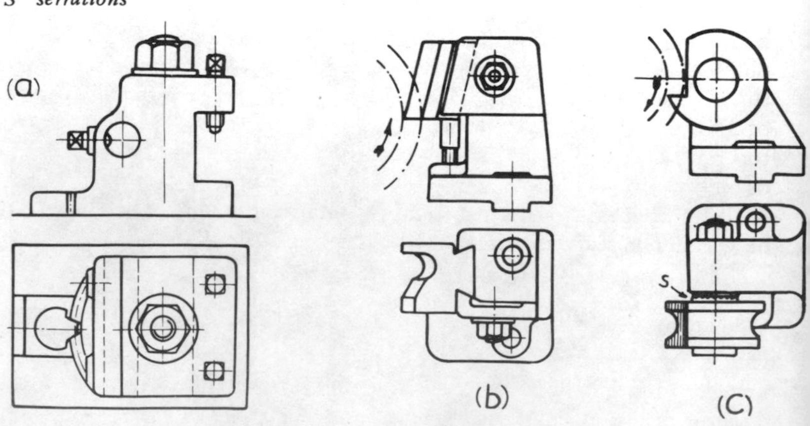

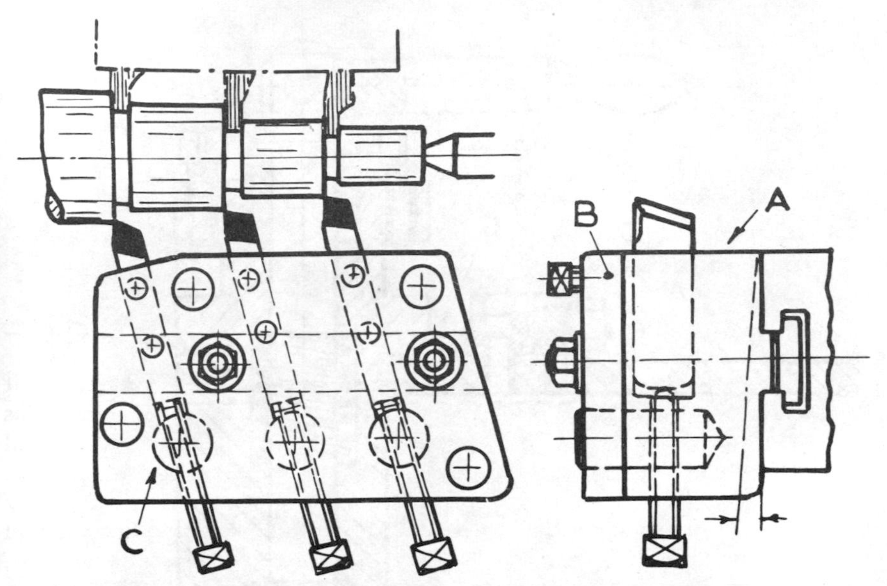

Figure 8.2.
A tool holder designed for multiple cutting operations:
A block B cap C plugs

the tool is set higher than the work centre to give clearance, and to prevent the tool from revolving on the holder a series of serrations S are milled on the tool to correspond with those on the holder.

There are patented tool blocks available in which the holder slots into a main unit. The holder can be removed and can later be replaced with the certainty that the tool will be in exactly the same position as before. Fine adjustment is available for accurate setting or to compensate for tool wear. These blocks are useful for batches of the same workpiece which tends to re-occur. For multiple cutting Figure 8.2 shows a useful design of tool holder. The block A has a cap B, while the tools, which are ground on the shanks, fit into slots at an angle of 75°. The three plugs C are fitted with adjusting screws, and, if the tools are required for machining steel, the base of the block may be machined at an angle of 5 to 8°, as shown by the chain lines. This feature automatically provides top rake on the tools. When grooving tools are used on a rear rest, these, or any other tool, can be removed or replaced without disturbing the other tools by slackening the screws and by removing a plug. The tool can then be withdrawn rearwards from the holder.

TYPES OF TURRETS

One of the most useful types of tool holders is the square, pentagon or hexagon turret. While formerly mainly restricted to lathes, the introduction of NC machine tools has extended the use of turrets to machines

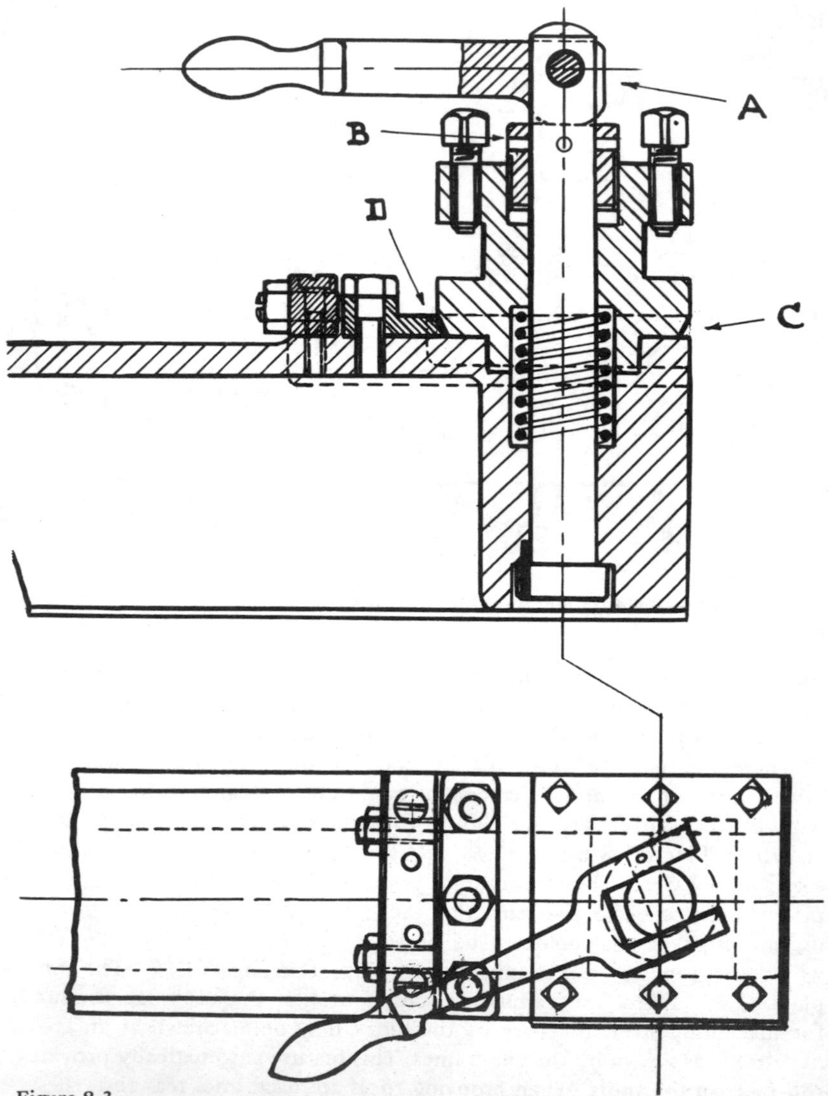

Figure 8.3.
A square turret tool holder for boring and turning mill:
A cam-shaped projection B nut C bevel in turret base D plate

for boring, drilling and milling operations where a machining sequence is required. Figure 8.3 shows a hand-operated square turret used on the side head of a boring and turning mill. Clamping is by the cam-shaped projection A on the end of the lever when in a horizontal position; release is obtained by bringing the lever to a vertical position, while the compression spring lifts the turret from the face of the slide. In this position it may be revolved to the next station and may be reclamped by bringing the lever down to its original position. Fine adjustment between lever and turret is obtained by the nut B. For locating purposes the base of the turret is bevelled on each of the four sides as at C and is held down in position against the plate D which can be adjusted as required. This arrangement provides a more accurate location than can generally be obtained by the use of taper bushes and a plunger.

TURRETS FOR NUMERICALLY CONTROLLED LATHES

To show how versatilely a turret can be tooled, Figure 8.4 shows the two indexing turrets on a NC lathe by Dean, Smith & Grace Ltd. On the front vertical rotating turret eight tool holders can be bolted onto hardened and ground locations. The turret is indexed electrically by the motor shown, and a flexible coolant pipe is fitted. The rear-mounted four-station turret rotates in a horizontal plane, again by an electric motor shown at the top. This turret is primarily for internal machining operations, a

Figure 8.4.
Indexing turrets fitted to an NC lathe
(By courtesy of Dean, Smith & Grace Ltd, Keighley)

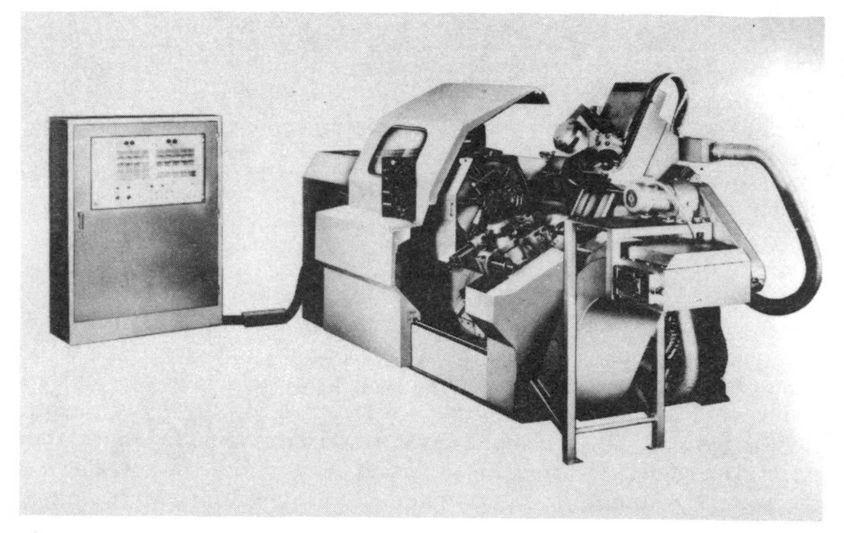

Figure 8.5.
Angular-set turrets on Wickman chucking lathe
(By courtesy of John Stirk & Sons Ltd, Halifax)

boring bar being shown in operation. Each station has its own flexible coolant supply pipe, and all supply points are controlled by switch or tape command.

The tooling arrangements for the Wickman chucking automatic lathe (John Stirk & Sons Ltd) are shown in Figure 8.5. The machine operates on an automatic cycle from a programmed control system in which both turret slides can operate independently. The turrets are indexed hydraulically, and there is a 24 in (600 mm) hydraulic chuck for work holding, while spindle speeds can be changed under load. The machine is driven by a 50 hp motor through six electro-magnetic clutches.

A feature of the bed construction is the 45° sloping face for the turrets, which, apart from providing convenient work-loading facilities, enables easy swarf removal. Thus, while a large clearway is provided in the bed to allow some cuttings to fall to the rear of the machine, the front section is fitted with a conveyor which leads swarf away from the base, up the incline and out onto a horizontal section to a removable bin or onto another conveyor if desired.

HOLDERS FOR BORING, DRILLING AND MILLING

While not having the complexity of turret tooling, the development of the machining centre has necessitated quick-acting tool holders for rapid replacement of tools. The British Motor Corporation uses chucks of their own design (Figure 8.6) for holding drills, reamers and taps. The chuck provides a positive drive, the construction comprises the two

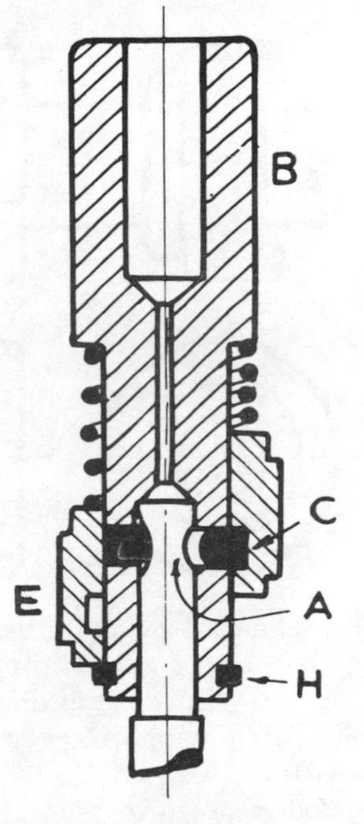

Figure 8.6.
A quick-change drill chuck:
A notches
B body
C driving keys
E sleeve
H spring clip
By courtesy of British Leyland

notches A, and the body B is provided with two opposed slots for the driving keys C. Driving is effected by the sleeve E being in the bottom position, left hand half, but when raised to the right hand position, the tool can be removed from the chuck. A spring holds the sleeve in the engaged position, the forward movement being limited by the spring clip H. A similar design is often used on broaching machines for connecting the broach to the drawhead.

Figure 8.7 shows a quick-acting arrangement for holding and driving short boring bars. Bars are stored in a circular disc attached to the machine, and, when swung in front of the spindle nose, a push forwards sends the bar into the spindle nose. A half-turn of the quick-acting nut C then locks it into position. After each operation, a similar half-turn allows the bar to be withdrawn into the bar-holding disc. The bars are located in the spindle nose by a parallel portion, while positive driving is obtained by means of a flatted part on the bar engaging a slot in the spindle end. Such devices are essential on machines with automatic tool-changing requirements where tools are transferred from machining to magazine-storing centres and vice versa.

ROLLER STEADY BOX TOOLS

A roller steady box tool is the most important tool holder used on capstan and turret lathes operating on bar work. Figure 8.8 shows the design which

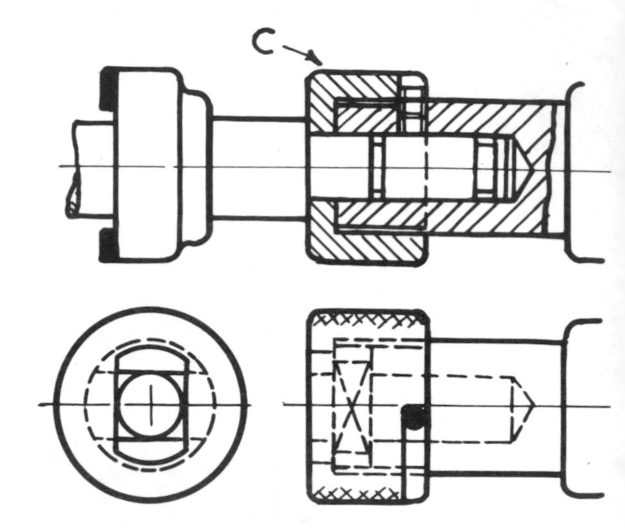

Figure 8.7.
A boring bar holder with quick-release facilities

comprises two rollers A mounted on roller bearings. The tool is bolted to the face of the hexagon turret and carries a turning tool Z set slightly in front of the rollers so that, when a bar of original diameter Y is reduced to X, the rollers immediately support the work and ensure adequate support as turning proceeds.

Adjustment of the rollers down to the bar is on the slides B, while ingress of cuttings is prevented by means of the plates D, and a hardened washer E is provided at the opposite side to take end thrust. By this type of tool holder a large reduction of bar diameter is possible by one cut, and a high work finish is obtained by the burnishing action of the rollers

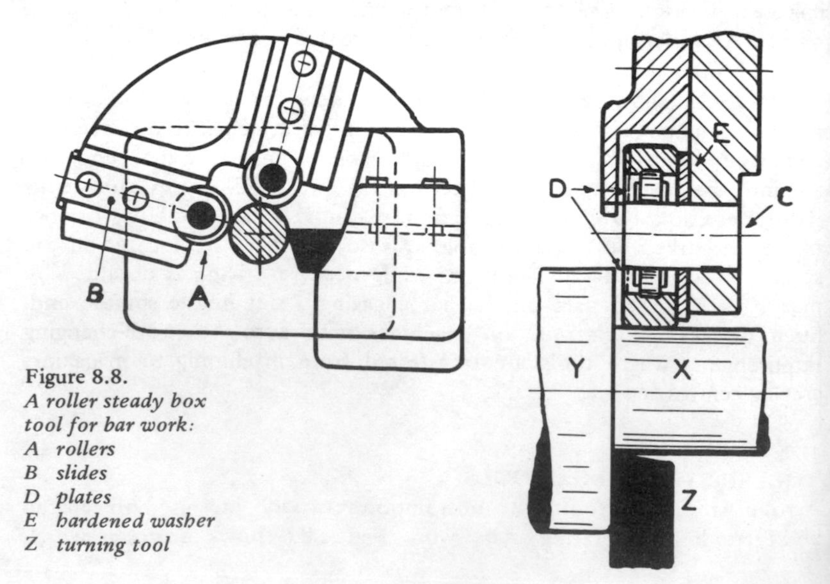

Figure 8.8.
A roller steady box tool for bar work:
A rollers
B slides
D plates
E hardened washer
Z turning tool

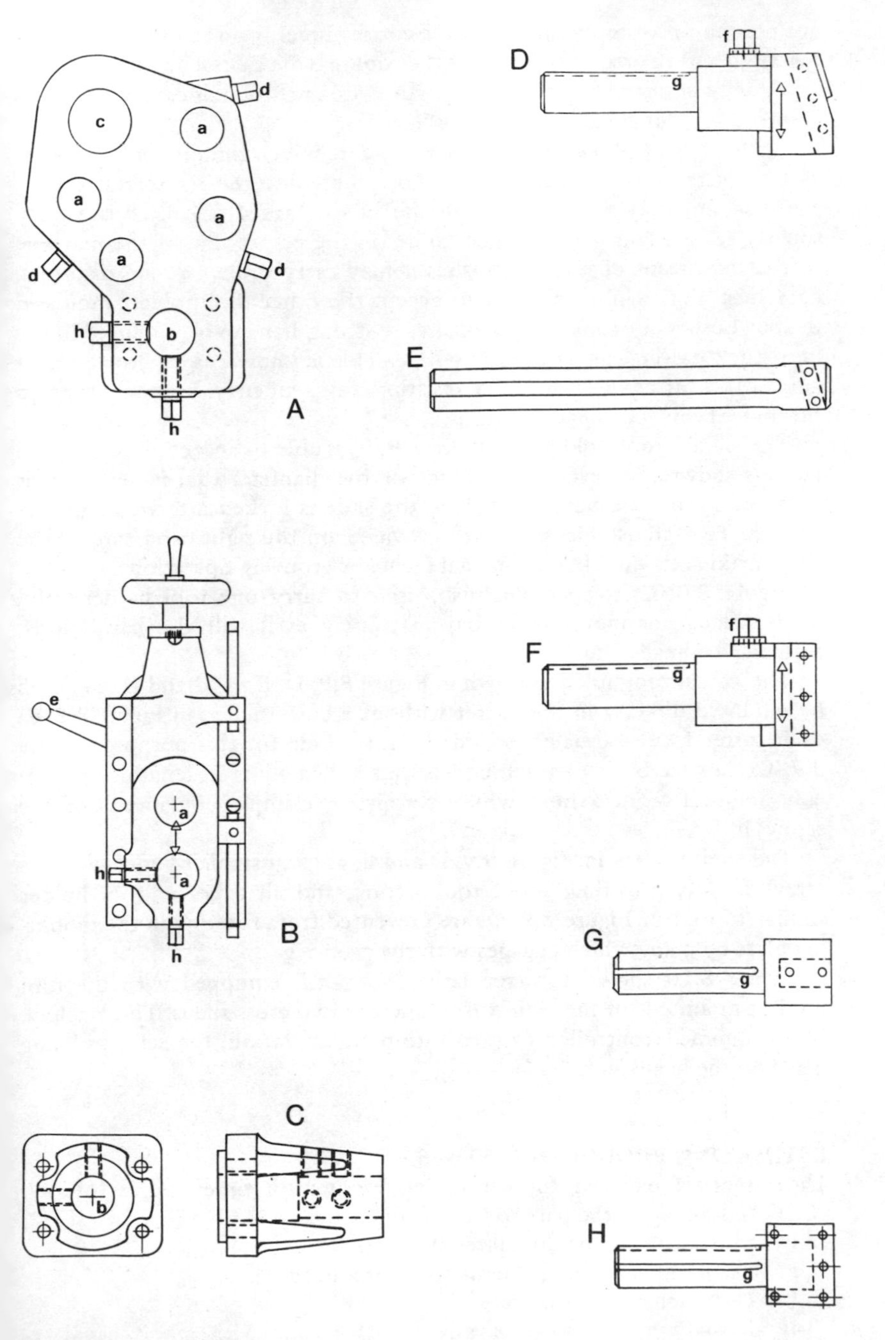

Figure 8.9.
Tool holders and supports for chucking work

against the pressure of the cut. Free-cutting steel is often used for the bar material, and the top rake angle of the tool may be as much as 25°.

Equally important, for chucking work, are the tool holders and supports shown diagrammatically in Figure 8.9.

Figure 8.9 A, shows a multi-hole support which mounts on the turret and can carry tool holders such as those shown on the right-hand side of the diagram; they are mounted in the holes marked a. Hole b is usually slightly larger, and it is arranged to be on the centre line of the machine so that by means of suitable bushes it may carry drills, reamers or boring bars such as E whilst still able to accept the other tool holders mounted in split bushes. For maximum rigidity a strong bar can be mounted in the headstock casting, aligned with hole c. This is known as a pilot bar as it guides the turret and assists in resisting any tendency for the turret to tilt under heavy cutting conditions.

The sliding tool holder in Figure 8.9, B, is able to accept all of the tool holders shown; however, adjustment of the diameter that is being cut is made easily by the handwheel, and the slide is locked afterwards by the lever e. Two adjustable stops are provided on the right-hand side of the slide, making it suitable for internal facing or grooving operations.

Figure 8.9, C, is a simple bush, used to carry one tool holder only; it also finds common use in bar machining as it will also carry drills, reamers, die heads, etc.

The actual tool holders shown in Figure 8.9, D, E, F, G and H, can each be mounted directly in the turret without a bush such as in Figure 8.9, C; each turret face is usually provided with a hole for this purpose. Figure 8.9, E, shows a boring bar which has flats milled along its length to permit easy removal from a hole whilst permitting clamping by means of the screws h.

The tool holders in Figure 8.9, D and F, are adjustable by means of the screw f, thus providing rapid tool setting, and all types of tool holder similar to that in Figure 8.9, D, are prevented from rotating in the mounting plate by a key which engages with the groove g.

Figure 8.10 shows a turret lathe (Minganti) equipped with the tool holders mounted on the main turret and the two cross slides. The machine is programmed controlled to give automatic operation, the settings being made on the headstock.

ESTIMATING PRODUCTION TIMES

The essentials required for estimating production times are as follows.

(1) A drawing of the part to be machined.

(2) A specification of the material.

(3) The number of components to be produced.

(4) The machine to be used.

If alternate machines are available, select one with ample capacity, for to use a machine just large enough may mean that slides are at their

Figure 8.10.
A turret lathe equipped with tool holders on turret and cross slides. Programme control is often added to a turret lathe to give automatic operation, as on this machine. A typical tool set-up is mounted on the turret and the two cross slides (By courtesy of Minganti, Bologna)

extreme limit and liable to produce vibrations by lack of support.

Having settled on a machine, a capacity chart giving details is required. These will always include the range of speeds and feeds available and the principal dimensions. Thus for a lathe, Figure 8.11 shows a simple diagram in which A is the length admitted between centres, B the diameter swung over the bed, C the diameter swung over the saddle, D the size of the tool holder admitted in the rest and E the size of hole through the spindle. For a machine with a hexagon turret, sizes should be given for the turret face, the size of the locating hole in the face, and the size and position of bolt holes for attaching tool holders. For a machine on which work is held on a table, say, a horizontal boring machine, the main dimensions would include the minimum and maximum distance of the spindle over the table top, the size of table and size of tee slots, and the length of the various traverse motions.

Figure 8.11.
A capacity chart for centre lathe

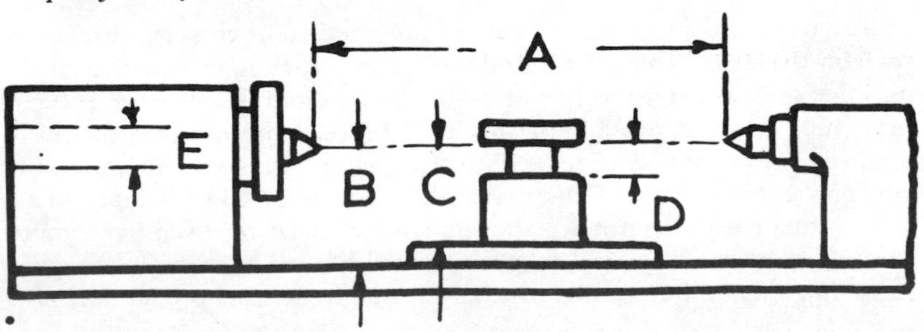

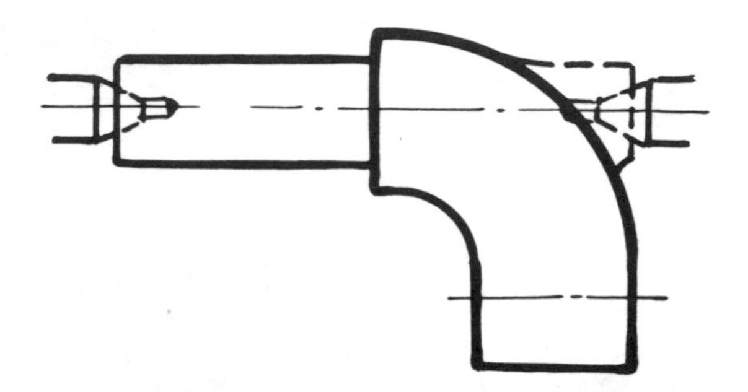

Figure 8.12.
An example of component simplified to ease machining

Production may be increased by using special tooling equipment, but this may prove expensive on a small batch of components; so consideration should be given to whether the use of standard equipment, slightly modified, is feasible. This may mean fitting special-shaped jaws to an otherwise standard chuck instead of a completely new fixture or adapting a standard machine vice with jaws suitable to grip a particular workpiece. In some cases lugs or pads on a casting will enable a difficult-shaped component to be held in a standard chuck, the lugs being removed after completion of the operation. A simple case is shown in Figure 8.12, where it is required to turn the spigot on the component which is almost impossible to hold in a chuck, but, by providing a small boss, shown as a broken line, by centring each end the component can be mounted between centres, and the spigot can be turned with ease.

SPEEDS AND FEEDS

The suitable cutting speeds and feeds for the use of different tools and materials are given in Chapter 1 and will serve as a general guide for turning, drilling and boring operations, while operations such as screwing, tapping and forming are considerably lower.

To translate metres per minute into revolutions per minute depending on work diameter we use

$$\text{spindle speed (rev/min)} = \frac{\text{cutting speed (m/min)}}{\text{work diameter(m)} \times \pi} \quad \text{or}$$

$$\text{cutting speed (m/min)} = \text{rev/min of spindle} \times \text{circumference of work (m)}$$

The term feed indicates the rate of longitudinal or cross traverse of the tool on the work. Thus, if the feed is 0.5 mm for each revolution of the machine spindle and it revolves at 200 rev/min, the tool will move 100 mm in 1 min. It is not possible to specify definitely the feed that should be used for any given operation as this depends upon the material, depth of cut, power and rigidity of the machine. Feed combined with depth of cut and cutting speed determines the amount of metal removed per minute. It is often more economical to prolong tool life and thus to save frequent regrinding than to put a strain on the machine to increase production.

OPERATION PLANNING OF BAR WORK

If we consider the tool set-up for producing the front wheel axle on a capstan lathe shown in Figure 8.13(a). The first operation for this component and others made from bar is to bring the bar out of the collet chuck and to feed it to the stop in the turret. The material is of mild steel 20 mm in diameter, and the cutting tool of high-speed steel operating at 38 m/min. The tool lay-out is shown in Figure 8.13(b), and Table 8.1 gives the operations and production times.

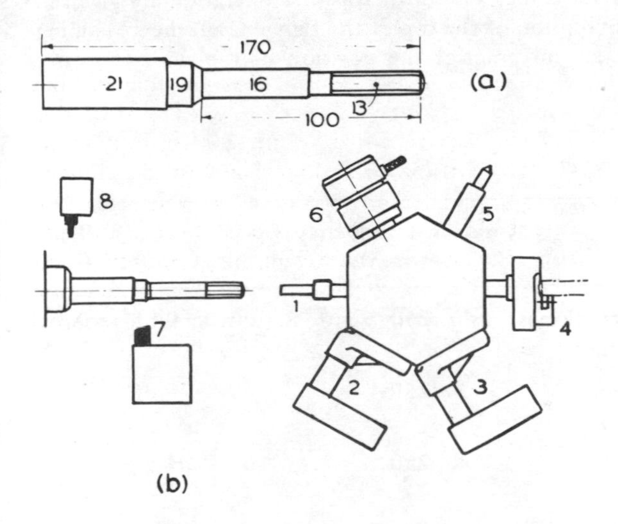

Figure 8.13.
A tool set-up for bar work on a capstan lathe

Table 8.1.

Operations	Travel (mm)	Feed (mm/rev)	Rotational frequency (rev/min)	Time (min)
(1) Feed out to stop in turret				
(2) Turn 13 mm diameter with box tool	50	0.025	550	0.58
(3) Turn 16 mm diameter with box tool	50	0.025	550	0.58
(4) Form end of bar	3	Hand	550	0.1
(5) Centre drill	—	Hand	550	0.25
(6) Cut thread with diehead	32	—	54	0.39
(7) Form and chamfer 19 mm diameter	—	Hand	550	0.15
(8) Part off with stepped tool	11	Hand	550	0.25
			Cutting time	2.30
			Operating time	0.40
			Contingencies	0.50
			Total time	3.20

OPERATION PLANNING OF CHUCK WORK

The component is a brass cover required to be machined all over (Figure 8.14). A turret lathe is used with the cover held in a three-jaw chuck as indicated in Figure 8.15. The tools are held in the hexagon and square turrets with the reference letters near the tools denoting the corresponding letters on the work surfaces to be machined. A finishing cut on the same lines as the roughing cut is required; so a face of the turret denoted by the figure 2 is reserved for this equipment shown by the broken lines.

The speeds, feeds and times taken for the first operation are given in Table 8.2. With the exception of the top of the flange C, all the machining is inside the casting, and in practice the position of tool C is vertically above the boring bar.

Table 8.2.
First setting

Operation	Rotational frequency (rev/min)	Speed (m/min)	Feed (mm/rev)	Time (min)
(1) Bring up 1 after chucking	250	60	0.05	0.30
Rough turn and bore				0.36
(2) Bring up 2	250	60	0.05	0.10
Finish turn and bore				0.36
(3) Bring up 3	250	60	Hand	0.20
Rough face				0.32
(4) Bring up 4	250	60	Hand	0.20
Finish face				0.32
Remove work				0.25
				2.41
For tool setting, changing and allowances 25%				0.6
Total Total time				3.01

For the second setting the work is held on the top of the machined flange in a three-jaw chuck with soft jaws, but, to prevent the casting from being squeezed out of concentricity owing to its frailness, it is first located on a spigot. This also locates it at the correct distance from the chuck and square with the front flange F. The operations are denoted by reference letters as before, and it will be seen that several tools are required for very minor operations such as chamfering corners, bevelling and undercutting for screwing. For this it is advantageous to leave plenty of length on the shank, because the first few threads are often imperfect, and these can then be parted off.

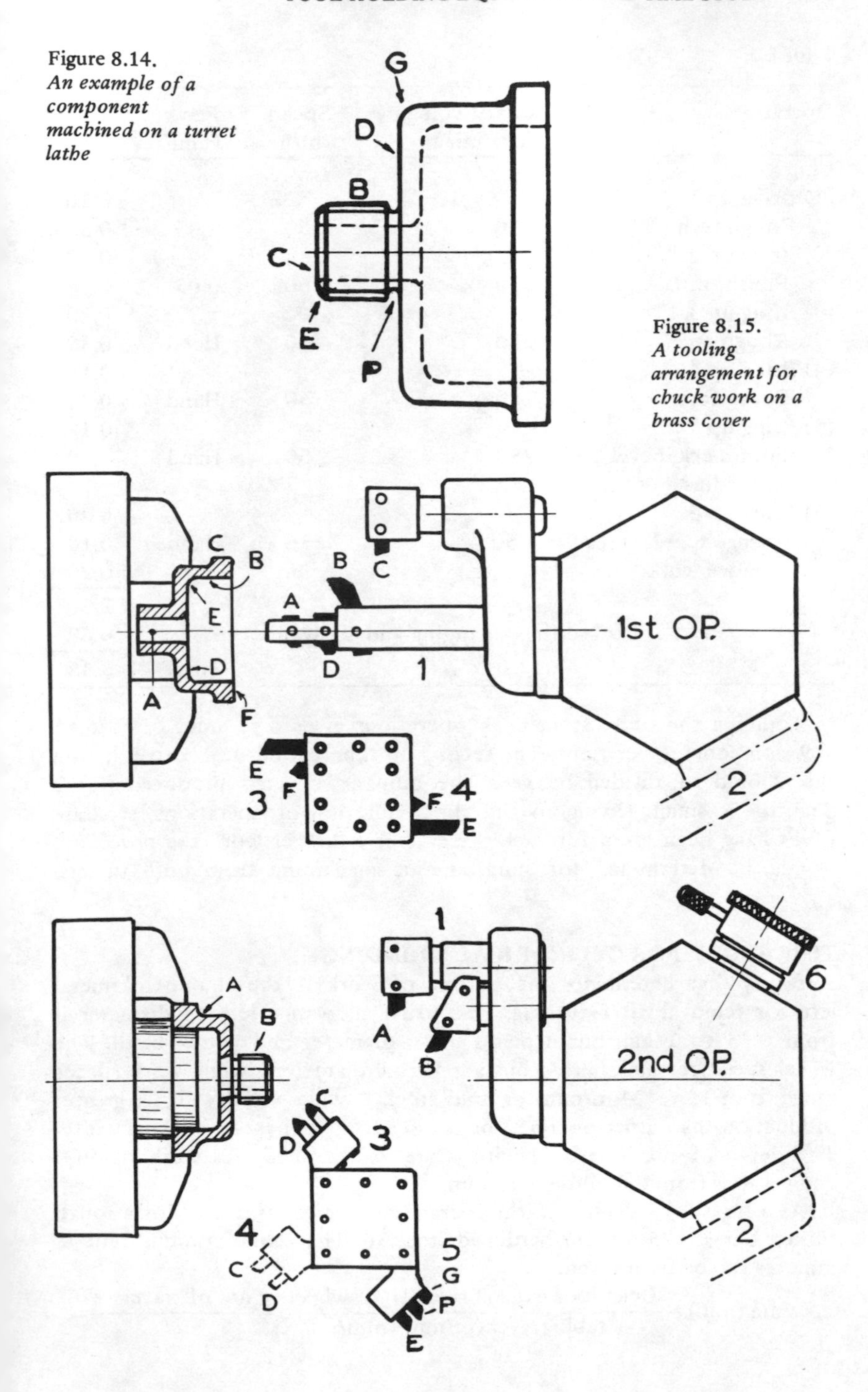

Figure 8.14.
An example of a component machined on a turret lathe

Figure 8.15.
A tooling arrangement for chuck work on a brass cover

Table 8.3

Operation	Rotational frequency (rev/min)	Speed (m/min)	Feed (mm/rev)	Time (min)
Chuck				0.2
(1) Bring up 1				0.10
Rough turn	250	60	0.05	0.36
(2) Bring up 2				0.10
Finish turn	250	60	0.05	0.36
(3) Bring up 3				0.20
Rough face	250	60	Hand	0.32
(4) Bring up 4				0.10
Finish face	250	60	Hand	0.32
(5) Bring up 5				0.10
Form neck, bevel and radius	250	60	Hand	0.10
(6) Bring up 6				0.20
Change speed; thread	150	36	Hand	0.10
Remove work				0.20
				2.76
Tool setting, changing and allowances 25%				0.69
			Total	3.45

Totalling the times from both operations gives 6.45 min, or 60/6.45 = 9 components per hour. The setting-up time would be 1½ to 2 h, and this should be divided between the number of parts produced if the quantity is small. Owing to the short duration of operations, separate times have been given for each turret, but on larger work the procedure should be determined for simultaneous machining from both turrets.

TIME STUDY FOR CYLINDRICAL GRINDING

A factor that determines the output of work is the amount of metal left for removal after turning. The usual allowance left on diameter is from 0.25 to 1 mm but depends upon diameter and work length. The wheel speed remains fairly constant for all diameters, while work speeds range from 15 to 24 m/min for mild steel. A wide wheel will give greater production than a narrow one, for the feed rate is based on wheel width. The depth of cut is variable with wheel width, feed and work rigidity, and it varies from 0.006 to 0.012 mm.

As a basis for calculation, the average number of traverses for a rough turned bar are 25 and for hardened steel 36. The actual grinding time in minutes can be found from

$$\text{time (min)} = \frac{2(\text{length of cut} + 2 \text{ breadth of wheel}) \times \text{no. of traverses}}{\text{table travel (in/min (mm/min)}}$$

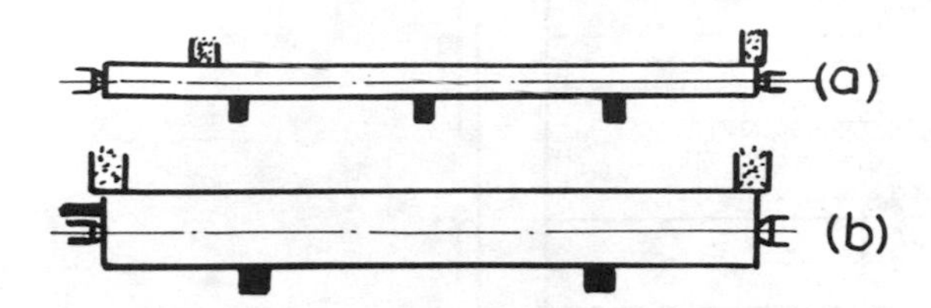

Figure 8.16.
Examples of cylindrical grinding on two components

The additional times to be added include the setting of work and stays, say, 2 min. each for 1 stay per foot of length, also a gauging time of 1 min for each diameter, 5 to 10% of grinding time for truing the wheel and other allowances. These figures are given as a general guide; to show the variable conditions that can be obtained two cases are given for the grinding of boring bars (Figure 8.16).

The case shown in Figure 8.16(a) necessitates considerable care because the work is liable to bend and vibrate, and this requires the support of three rests. Additional time is taken up in accurate setting, for the bar which is case hardened is limited in the depth of metal that can be removed. A further complication arises in that the bar is of one diameter only, so that it cannot be traversed ground for the full length but requires changing end to end for roughing and finishing cuts.

A small wheel 300 mm in diameter with a 15 mm face would be suitable with a work surface speed of 10 m/min roughing and 15 m/min for finishing. The depth of roughing cut is 0.025 mm and 0.012 mm with a finishing cut. The traverse rate is 800 mm/min.

In contrast the case shown in Figure 8.16(b) presents no difficulty. The large diameter enables a driving pin to be fitted in the end, so that the wheel traverse is from end to end. A wheel 600 mm in diameter with a 60 mm face could be used with a work surface speed of 15 m/min for roughing and 25 m/min finishing. The traverse rate is 1,500 mm/min. Two steady rests are required.

The time taken for grinding a shaft of several short lengths is increased not only by the change in diameters but by the period of dwell which takes place at the end of each traverse. On the other hand, a short length on a rigid shaft gives an opportunity to use a wide wheel for plunge grinding.

The work speed may be higher for internal grinding than for external work, but the dressing of the grinding wheel absorbs 10% of the cutting time. Small wheels have a short life, and 5% of the cutting time is a suitable allowance for wheel changing. The traverse per revolution of the work may be up to three-quarters of the wheel width with an average in-feed of 0.0025 mm for holes below 15 mm in diameter, 0.004 mm for holes from 15 to 30 mm in diameter and 0.005 mm for larger bores. Dressing time can be reduced to a few seconds by use of a diamond roller, as shown in Chapter 6.

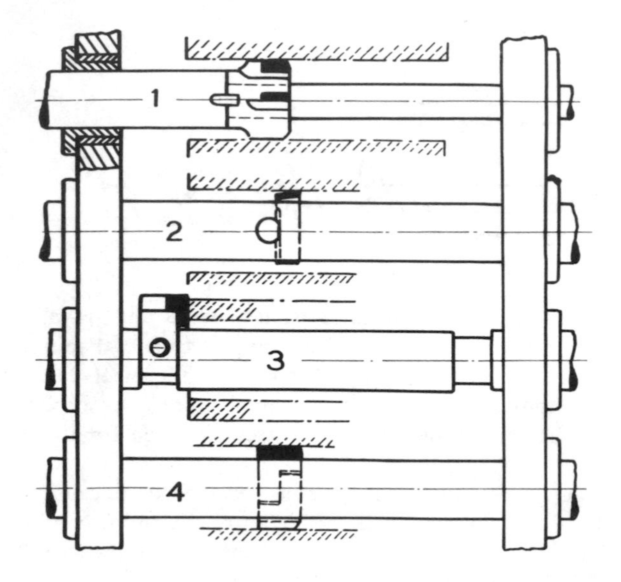

Figure 8.17.
The tooling arrangements for boring operations on lathe tailstock

BORING AND FACING OPERATIONS

If we select the machining of a lathe tailstock as an example, the sequence of operations is shown in Figure 8.17. The casting is located in a jig with end brackets to carry guide bushes for the boring bars. Under these conditions it is preferable that the bars be connected to the boring machine spindle by flexible couplings.

FIRST OPERATION

Rough with four-flute core drill to remove metal quickly.

SECOND OPERATION

Semi-finish bore to 0.140 mm by using a single-point tool to straighten the hole. A cutting speed of about 30 m/min is usual for cast iron, and feed rates of 0.75 mm/rev for roughing and 1.5 mm/rev for finishing.

THIRD OPERATION

Face the end of the bore. It should be noted that the bar is grooved and that the cutter head does not fit all around the bar. This enables the facing head to be bolted in position without removing the bar from the jig, which is a time saver.

FOURTH OPERATION

The last operation is reaming. A floating double cutter as shown or a reamer is used to finish the bore. The cutting speed should be 6 m/min for cast iron or 10 m/min for mild steel. The feed rate should be from 0.8 to 5 mm/rev according to the hole diameter.

9

CUTTING FLUIDS AND SUPPLY OF COOLANTS - SURFACE TEXTURE

The function of a fluid for the assistance in machining metals may be described as follows.

(1) To cool the work and cutter.

(2) To wash away chips.

(3) To lubricate the bearing formed between chip and lip of the cutting tool.

(4) To prevent welding of work particles on the tool tip.

(5) To enable the cutting tool to provide a good surface finish on the work.

(6) To protect the finished product from corrosion.

Of these functions the cooling action is the most important. During every cutting action, heat is generated between the tool and work, and, if provision is not made for the removal of this heat, the temperature may become so excessive that the cutting edge of the tool breaks down. There is in addition the possibility that heat will expand the work during machining, so that work measured at this stage may be found undersized when it contracts.

In deep hole drilling the ability of the cutting compound to wash away cuttings is of importance. The chips produced are broken into small pieces, but the tendency to pack together is such that very high pressure is required to clear them, the delivery rate and pressure reaching 225 l/min and 8.75 kg/cm^2 respectively (see Figure 9.8).

In the machining of cast iron, aluminium, high-carbon steel and some grades of brass, the lubrication action is of little importance, as the chips break into small pieces, but it is very important when machining materials such as low-carbon steels, where long chips are produced that curl back over the tool. In such cases a bearing is produced in which the frictional resistance is severe, and, unless the cutting compound is a lubricant as well as a coolant, friction will result in rapidly wearing out the tool..

CLASSIFICATION OF CUTTING FLUIDS

Under *aqueous fluids* we classify water and aqueous solutions. The *neat cutting oils* include mineral oils, mineral-lard (compounded) oils, fatty oils and extreme pressure (EP) oils. These are also *aqueous emulsions of soluble cutting oils.*

Water has unique potential cooling power, but it rapidly rusts steel. Soda and soap solutions have been used and are an improvement but are now little used except occasionally for grinding. To obtain improvements in lubrication it has been necessary to turn to oils; mineral oils are used, but fatty oils are more effective. Compounded oils generally known as mineral-lard oils containing up to 30% of fatty oil are quite efficient. Oils cannot cool as well as aqueous cutting fluids do, because their specific and latent heats are lower, but their greater lubricating power reduces heat generation and compensates for their inferiority as coolants. Lubricating power also helps to give a better workpiece finish and prolongs tool life.

The increase in machining speeds and development of tougher materials caused the introduction of the EP cutting oils. The additives enable them to prevent the welding of chip to tool that tends to take place when tough materials are being machined. The most important elements are sulphur and chlorine. Their action is not confined to anti-welding; they also help to lower friction and to improve wetting power.

SOLUBLE CUTTING OILS

These consist of mineral oil plus an emulsifying agent to disperse the oil into the water and so to form an emulsion. The term soluble oil is really a misnomer, since it implies the formation of a true solution, instead of the colloidal dispersion that is actually formed. The oils do not appear to dissolve in water, however, and the dispersions are referred to as emulsions.

The soluble oil enables use to be made of the cooling power of water without its disadvantages, i.e. corrosivity and poor wetting power. Since a soluble oil emulsion is nearly all water, its cooling power is high, but the oil retards the corrosive action of water and gives a certain amount of lubricating power.

OPAQUE AND CLEAR SOLUBLE OILS

Most soluble oils form opaque milky emulsions, the milky appearance being caused by the scattering of light by the small dispersed particles of oil. This has the disadvantage of preventing a clear view of the tool and workpiece. On the other hand, clear soluble oils form translucent emulsions through which is possible to see the workpiece. The reason why the emulsion is clear is that the oil particles are not large enough to scatter light, while other advantages include good anti-rusting properties and long service life, and also for grinding operations a minimum tendency to clog the wheel.

SELECTION OF CUTTING FLUIDS

HIGH-CARBON STEELS

Because high-carbon steels tend to soften at temperatures as low as 200°C, this heat generation must be kept as low as possible. Soluble oil is the best for this purpose, for there is no danger of welding during the machining operation.

HIGH -SPEED STEEL

A wide range of cutting fluids may be used, the choice being determined by the workpiece material and the nature of the operation. In general, soluble oil is satisfactory.

STELLITE CEMENTED CARBIDE AND DIAMOND

These types of cutting tools are generally used without cutting fluids, for the risk of welding is very slight. When a cutting fluid is used, its purpose is to wash away chips and to prevent work distortion; soluble oil is excellent.

INFLUENCE OF WORKPIECE MATERIAL

MAGNESIUM AND ITS ALLOYS

Magnesium and its alloys develop such a low cutting pressure that a coolant is not necessary, and they are generally machined dry. Aqueous emulsions are prohibited owing to fire risk, which exists because of the combustibility of the metal. When fluids are used, neat oil is the best, for light oils only are satisfactory.

ALUMINIUM AND ITS ALLOYS

Again, dry machining is usual, but light mineral oil can be used, while soluble oil emulsions are sometimes used.

YELLOW METALS

Soft brasses can be machined dry. 'Active' EP oils should not be used because of the free sulphur which causes staining of yellow metals.

COPPER

This is a difficult metal to machine because of its tendency to 'drag'. Mineral-lard or 'inactive' EP oils are used.

CARBON AND ALLOY STEELS

All types of cutting fluids are used, the choice depending on the tensile strength of the steel, the severity of the operation and the surface finish required. For mild and free-cutting steels, emulsions and mineral-lard oils are adequate, but 'active' EP grades are necessary for the higher-tensile varieties.

STAINLESS AND HEAT-RESISTING STEELS
Austenitic stainless steels are difficult to machine, for the work hardens under the cutting action. Powerful EP oils of the 'active' type are necessary for all these materials.

CAST IRON
The cutting action produces brittle chips, and machining is generally performed dry. The heat-resistant cast irons should be treated like heat-resistant steel.

TOOL WEAR
Neat oils give less tool wear than emulsions do; consequently, if great accuracy is required, neat oil should be used. In work such as broaching, milling and forming operations where tools are expensive, the saving in tool costs off-sets the higher cost of neat oils. Good tool life on automatic lathes is necessary to ensure long runs without tool grinding; so again neat oil should be used.

BROACHING
This can be a severe machining operation, and in such cases powerful EP oils are advisable. For the simpler operations such as keyway cutting, emulsions are satisfactory.

DRILLING
For ordinary operations, cast iron can be drilled dry, while soluble oil is suitable for mild and medium steel. For very hard steel, neat light or soluble oils are used. For deep hole drilling neat oils are generally chosen

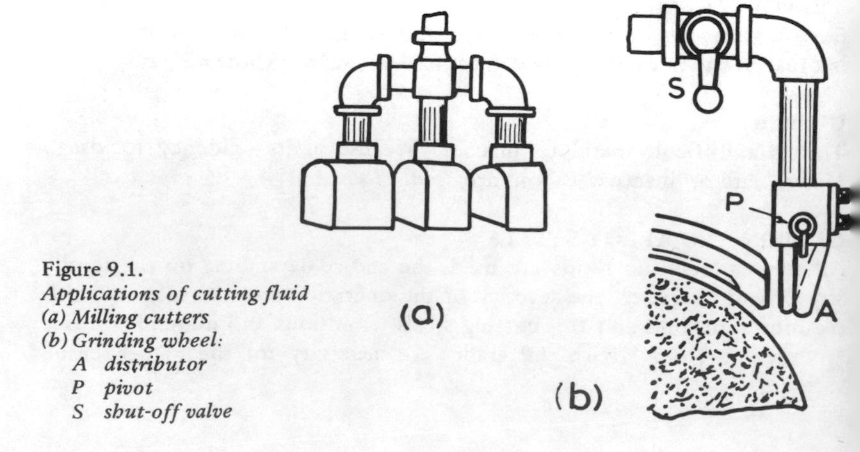

Figure 9.1.
Applications of cutting fluid
(a) Milling cutters
(b) Grinding wheel:
A distributor
P pivot
S shut-off valve

because of the severe friction developed. They must be able to wash away chips and must be applied at high pressure.

AUTOMATIC LATHES
Because of the high cost of the machines, it is of importance to keep them free of rust; so neat oils are preferred, even with carbide tools.

PLANING AND SHAPING
Lubricants are rarely used, for much of the work is done on cast iron.

GRINDING
Grinding is a mild operation, but cooling is hampered by the poor heat conductivity of the grinding wheel. Thus emulsions with their high cooling power are almost invariably used. Neat oils are sometimes used to give an improved surface finish or for grinding hardened surfaces such as gear teeth to reduce the risk of cracking from the quenching action.

THREAD GRINDING
This is a severe operation, for longer cuts are taken than in ordinary grinding. Moreover, the low area of contact leads to the setting up of very high stresses and pressures. Because of this, powerful EP oils only should be used.

APPLICATION OF CUTTING FLUIDS

Figures 9.1(a) and (b) show applications of the way cutting fluids are applied to the workpiece, showing the difference in the nozzles and the variation between clear oil and emulsion. The rate of flow is varied for different conditions, an average figure being 3 to 5 gal/min (13.6 to 22.7 l/min), for each tool. Only a small proportion of this reaches the actual point of cutting, the remainder inundating and cooling the tool shank, workpiece and adjacent parts. Sometimes the rate of flow may be more than double that given, particularly on grinding machines with multi-wheels. The sump capacity varies from 20 to 150 gal (91 to 781 l) depending on the size and type of the machine. As a rule about 10% replacement of fluid is required after a day's work owing to loss by volatization, to loss on chips and to fluid wasted on the machine parts.

A full flow of liquid around the tool and workpiece to remove heat from the tool, chip, work and machine bed is required and should commence before cutting is begun.

Although most machine tools each have their separate supply of coolant, centralized supply systems supplying batches of machines are in use, but only when numbers of machines require the same coolant. The fluid from the machines is pumped back to the main feed tank and is filtered before further use.

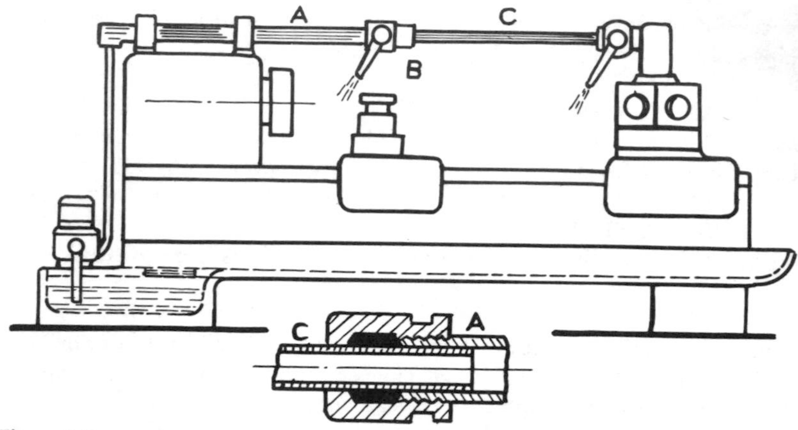

Figure 9.2.
A method of coolant supply to tools in square and hexagon turrets:
A pipe *B stuffing box* *C smaller pipe*

SUPPLY OF COOLANT

One of the methods of delivering coolant from a pump to a cutting tool is by means of flexible tubing, but it often happens, particularly on lathes, that the saddle has to travel a considerable distance on the bed, and telescopic tubing is required. This method is shown in Figure 9.2, where coolant is required to be available for the tools in the square turret as well as for those in the hexagon turret. The motor-driven pump supplies coolant to a pipe A supported by brackets on the headstock. There is a stuffing box B, shown detailed, and fitting into pipe A is a smaller pipe C which slides through the stuffing box as the hexagon turret saddle traverses

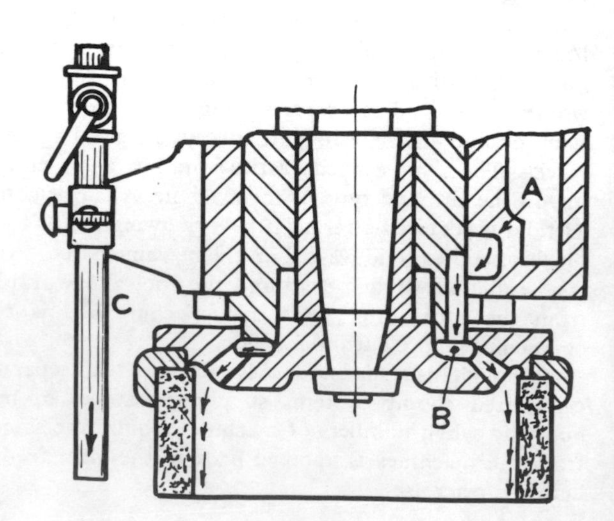

Figure 9.3.
A surface-grinding wheel mounting with coolant supply:
A passage
B annular recess
C auxiliary pipe

along. This smaller pipe is supported in a bracket on top of the hexagon turret, so that coolant supply is available from both pipes to the cutting tools. A single nozzle controlled by a tap is the simplest arrangement, but for the supply of fluid to, say, a set of gang milling cutters an arrangement such as that of Figure 9.1(a) controls the flow over each cutter.

Grinding operations require a large coolant supply so that a broad nozzle as in Figure 9.1(b) is provided to cover the full width of the grinding wheel. A distributor A may be adjusted vertically on the pipe and swung about pivot P so that it can be set at any angle. A shut-off valve is fitted at S.

A system used on a vertical surface grinding machine with a rotary work table is shown in Figure 9.3. From the pump a pipe connects to the wheel head and thence to passage A, from which the fluid passes down the annular recess B in the faceplate and then outwards and downwards inside the grinding wheel. The fluid in passing through the inclined holes is whirled at wheel speed and issues under the cutting face with considerable force. Thus the fluid not only cools but cleans the wheel face. An auxiliary pipe C is also available to deliver a heavy stream of fluid over the workpiece and table.

In Figure 9.4 the length of travel of a milling machine table makes it difficult to drain the table from both ends by means of tubing, for a long flexible tube from end A tends to be in the way of the operator. A better method is to drain the table by means of a pipe B which connects both right and left hand sumps on the ends of the table. This simplifies the arrangement, for pipe C is adjacent to the tank. Some of the coolant drains to the sump by way of grooves G, but chips often clog these and the tee slots; hence the use of the pipe B to get the coolant back to the tank.

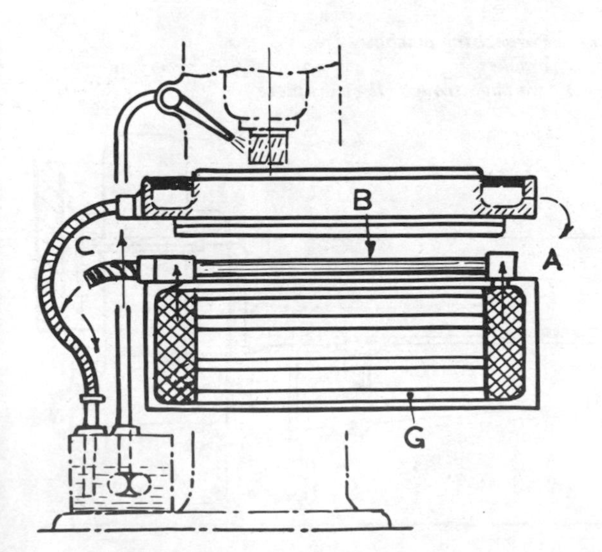

Figure 9.4.
A piping and coolant supply to a vertical milling machine:
A end
B and C pipes
G grooves

BROACHING OPERATIONS

A comprehensive system of coolant supply and drainage built into the design of a broaching machine is shown in Figure 9.5. The operation requires a considerable supply of coolant to wash away the small chips as well as to lubricate the cutting action of a very expensive cutting tool; in this case a broach to cut holes of dimensions 7 in x 7 in (178mm x 178 mm) on army tank components. The pump A is driven by vee ropes from the end of the driving shaft as shown and draws fluid from the reservoir R formed in the main casting of the machine. Connecting piping held by steel clips connects to the flexible metallic tube B and thence to a pipe and tap connections held in a bracket C on the faceplate of the machine to which the workpiece F to be broached is held.

A drainage trough D catches some of the used coolants and chips; the latter is held on a perforated plate recessed into the casting, while the fluid passes into the pocket and thence by the sloping return pipe to the tank. However, the flow of fluid is so great that a considerable quantity falls into the bed of the machine which is designed with a sloping base to conduct the fluid into the sump E, again fitted with a perforated plate, and thence by means of a vertical pipe which joins the main return pipe to the tank.

It is advisable to fit an inspection plate on the side of the sump for cleaning out. While strainers will keep the main cuttings out of the tank, in the course of time sediment builds up and requires removal. Cutting fluid as supplied is sterile, but it may pick up bacteria during use and may cause infection; thus the tank should be emptied periodically and should be thoroughly cleaned out. The procedure is to pump the tank as dry as possible and then to fill up with a 5% solution of caustic soda and to

Figure 9.5.
A cutting fluid arrangement for broaching machine:

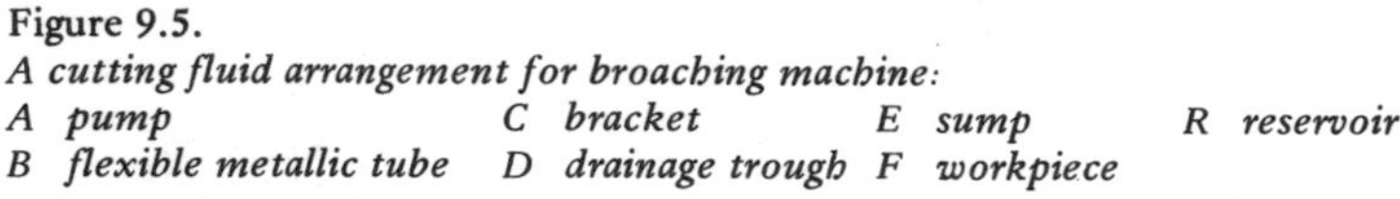
A pump *C bracket* *E sump* *R reservoir*
B flexible metallic tube *D drainage trough* *F workpiece*

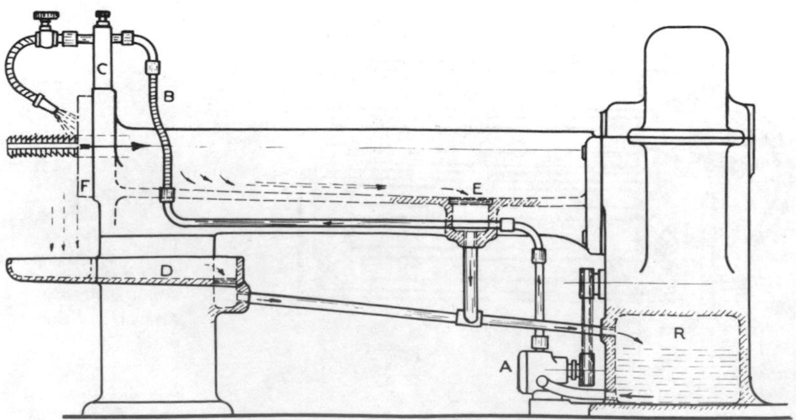

circulate for 10 minutes. Drain this away and refill with clean water, recirculate and finally replace with cutting fluid.

DEEP HOLE BORING

The production of precision holes with high ratios of length to diameter is accomplished by deep hole boring, which is a fundamentally different process from the use of an ordinary twist drill. Apart from the use of the process for rifle barrel production and cannon boring, long holes are required in the spindles of lathes and sometimes in milling machines. The difficulty of the operation arises from the length of the bore and the problem of removing chips, so that cutting fluid at high pressure is required together with good lubricating properties.

The Sandvik system is shown in Figure 9.6 where two boring tubes are used. The cutting fluid is introduced into the part connecting to the boring machine and flows between the supporting outer tube B and the inner tube A as far as the boring head C. Part of it is diverted through the annular nozzle D and straight back through the inner tube E. This sets up a partial vacuum in the chip passages of the drill body so that the portion of the cutting fluid flowing through the drilled hole F passes over the cutting edge, collects the chips, is sucked through the inner tube and is carried away. The cutting fluid serves both as a coolant transporter of chips and as a lubricator of edges and pads.

For a 50 mm (2 in) bore the oil pressure is 10.3 bar (151 lb/in^2), and the oil flow rate is 120 l/min (26 gal/min). The cutting heads for diameters from 20 to 65 mm (0.78 to 2.5 in) are intended as 'throw-away' units to replace with a new head when worn out. With the larger sizes of head, simply replace the carbide cutting tips and bearing pads, for the body section is retained.

Three other systems are in common use.

(1) The *GUNDRILL* method is in some ways similar to the Sandvik

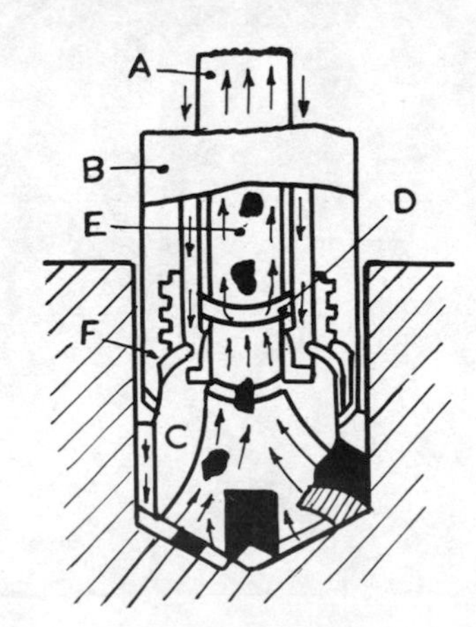

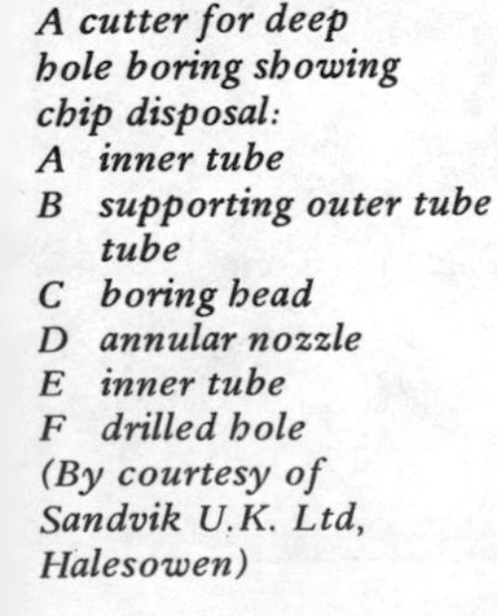
Figure 9.6.
A cutter for deep hole boring showing chip disposal:
A inner tube
B supporting outer tube tube
C boring head
D annular nozzle
E inner tube
F drilled hole
(By courtesy of Sandvik U.K. Ltd, Halesowen)

system in that cutting oil is carried within the drill to reach the point, but the drill shaft has a flattened area along its length which carries cutting oil and chips back from the point.

(2) The *B.T.A.* system delivers coolant to the point down the outside

Figure 9.7.
A Sass Deepmatic machine drilling holes in heat exchanger plates 1 metre deep. This is a modern deep hole drilling machine which can drill three holes simultaneously up to 40 in (1000 mm) deep. A typical vertical travel of the drilling head is 3 m, and the column may be mounted on a bed and sliding up to 10 m to permit a workpiece to be mounted whilst another is being drilled. The machine has an NC system (By courtesy of Comau Industriale, Turin)

of the drill shank which provides excellent lubrication for the drill and prevents chips from being trapped between the drill and the workpiece as they are returned with the cutting oil along the tubular interior of the drill. B.T.A. drills are supplied with cutting oil through some special guide bushes which embody seals pressed against the face of the component; in this case the component face must be smoothly machined prior to drilling.

(3) The *MECANO* drill is spiral fluted as a conventional drill, and two spiral holes are provided along the length of the drill to supply cutting oil to the two cutting points. Cutting oil and chips are returned along the spiral flutes.

The Sass Deepmatic machine (Figure 9.7), when drilling holes in heat exchanger plates, may be called upon to make up to 10,000 holes in any one plate up to a metre in depth. Extreme precautions must be taken to ensure an absolutely clean supply of cutting oil, as a failure to do this would result in severe damage to the high-pressure pumps in a very short time. Equally, scoring of the finished hole would occur with the Mecano or Gundrill systems if contaminated coolant is used. The manufacturers of this machine have provided extensive swarf conveyors, settling tanks and paper filtration with a capacity of 650 l/min to ensure that no chip particles of larger than 20 μm remain in suspension. It is interesting to note that on the three-spindle versions of this machine six pump units are used with drive motors totalling 120 hp to pump the cutting oil at up to 70 atmospheres of pressure. The machine has an NC system.

The pressure required for various sizes of hole up to 50 mm (2 in) in diameter is shown in Figure 9.8 and also of the appropriate flow rate of the coolant; curve P is the plot pressure against hole size, whilst curve F is the flow in litres per minute against hole size.

Figure 9.8.
Curves showing oil flow and pressure in deep hole drilling
(By courtesy of Comau Industriale, Turin)

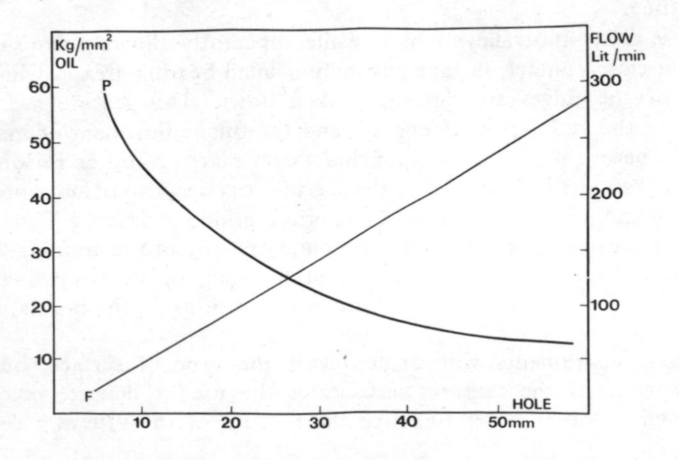

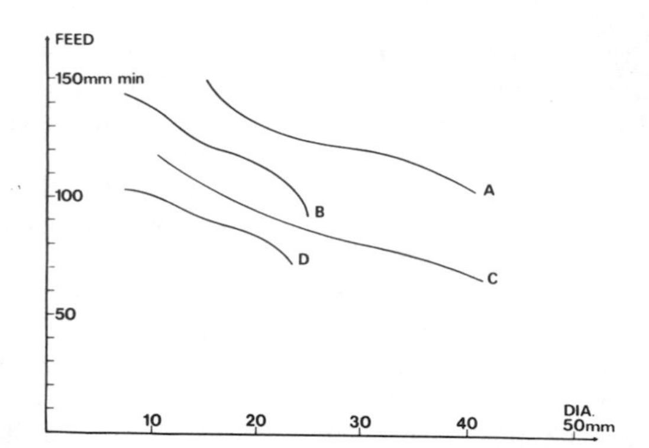

Figure 9.9.
Drill penetration rates in deep hole drilling
(By courtesy of Comau Industriale, Turin)

A simple calculation from the graph in Figure 9.9 will show the volume of chips to be removed from the cutting oil. The curves show drill penetration as follows: Curve A, mild steel, B.T.A. drill; curve B, mild steel, Gundrill; curve C, stainless steel, B.T.A. drill; curve D, stainless steel, Gundrill.

If we take curve A and assume a hole diameter of 25 mm (1 in), then each spindle will remove 61.35 cm^3 of mild steel per minute and thus a total of 184.5 cm^3 (11.7 in^3) for the three spindles on the machine.

SURFACE TEXTURE MEASUREMENT

Manufacturing processes involving metal cutting tend to leave on the surface of the workpiece characteristic patterns of hills and valleys known as the texture. This texture is deemed to have components of roughness and waviness where the height is small compared with the spacing of the crests and varies from a few thousandths to a few millionths of an inch. This depends upon the machining operation, and in practice when the surface irregularities are seen to have a marked and definite direction the measurement of a single cross section at right angles to their length will suffice.

These crests and valleys which, while apparently allowing two surfaces to be in close contact, in fact give only a small bearing area in which the peaks of the ridges are quickly broken down. This feature has made necessary the running-in of engines and the final adjustment of machine bearings before proper bedding-in has taken place. Another reason why these peaks must be removed is that certain operations not only produce the hills and valleys but, like commerical grinding, leave a surface of amorphous carbon formed by the heat and pressure of the grinding wheel. There are, of course, certain precision finishing operations which will produce very fine textures, but these are additions to the normal metal removal.

Optical instruments will often reveal the type of surface, but the measurement of the texture necessitates the use of delicate recording instruments, for in order to reach the bottom of the valleys a contact

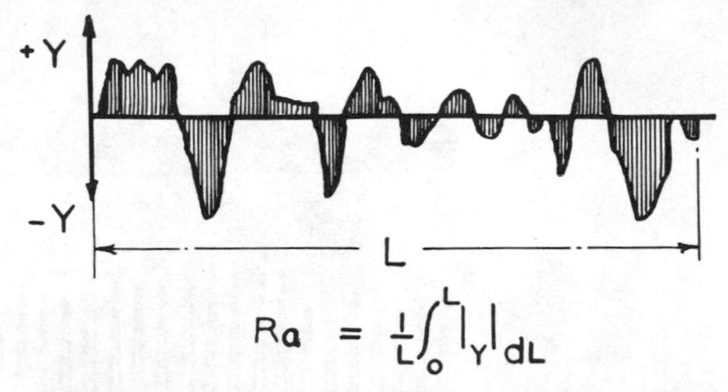

Figure 9.10.
The measuring parameter for surface texture

point of 0.001 in (0.002 mm) radius is required with a very light load in order not to damage the ridges. The B.S.I. standard (B.S. 1134) has been revised and is now in three sections: *Part I. Methods and Instrumentation; 2. General Information and Guidance; 3. Calibration of Stylus Instruments.* The parameters and tolerances are in SI units. Two measuring parameters are used; the main one is designated as R_a is identical with the previously known centre line average (CLA) (Figure 9.10). A second parameter gives a measure of the average peak-to-valley height of surface irregularities and is designated R_z. A term known as method divergence has been introduced for the first time, the need for which is as follows. The roughness average R_a is determined by integrating the ordinates of the profile with regard to a mean line which can be drawn as a straight line on the profile graph.

There is now international agreement about some symbols, e.g. V_v for vertical magnification, V_h for horizontal magnification and B_{max} for the meter cut-off. The international standard for surface texture is based on the arithmetical average height index R_a, but while this parameter is valuable it takes little account of the openness of the texture. An additional parameter, the average wavelength λ a has been defined and will be useful in the measurement of wear and to supplement the information given by R_a.

STYLUS SURFACE-MEASURING INSTRUMENTS

The peak-to-valley heights of the texture of a component may range from about 0.05 μm for fine lapped, through 1 to 10 μm for ground and up to 50 μm for rough machined surfaces. Peak spacings along the surface may range from 0.5 to 5 mm. Because of the need for portraying on a profile graph a sufficient length of surface, it is necessary to use a far greater vertical than horizontal magnification. This is shown in Figure 9.11(a) showing commercial grinding with a vertical magnification of 20,000 to 1 and 200 to 1 for a horizontal magnification. The average height of the surface is 0.61 μm. The example is from a diagram obtained on a Rank, Taylor-Hobson instrument, while further examples are from a Tomlinson recorder, indicating the surface texture of a steel bush in various stages of manufacture where Figure 9.11(b) is after turning, Figure 9.11(c) is after

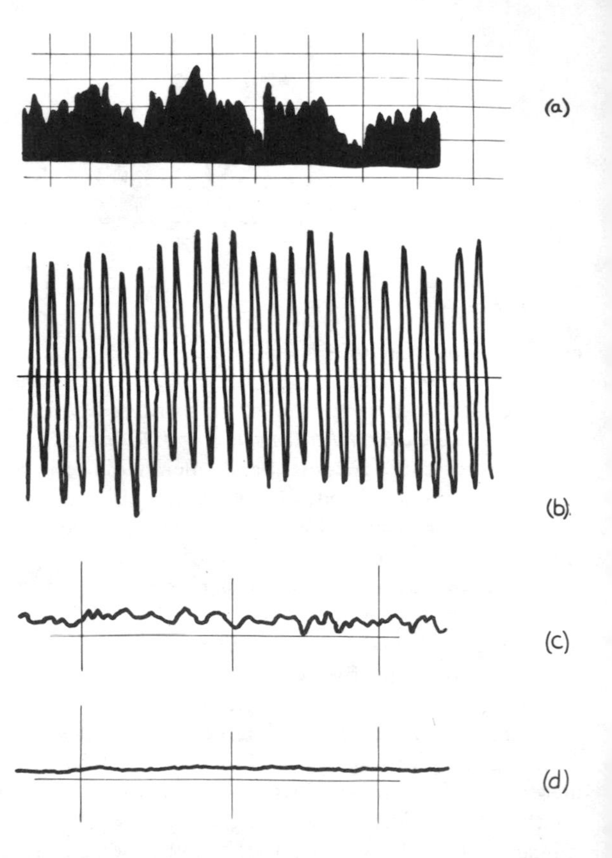

Figure 9.11.
(a) Chart of surface finish from shaft grinding
(b) to (d) Turning and finishing steel bush

grinding and Figure 9.11(d) indicates the smooth surface that can be obtained by the precision operation of superfinishing. The scale of these diagrams is 50,000 on the vertical scale and 50 on the horizontal scale.

Figure 9.12(a) shows a stylus instrument in which the housing P is hinged to the driving mechanism at H and is provided with a pair of rounded feet, or skids S; at least one of the feet rests on the work specimen W and slides across it with the stylus from A to B. Q represents a displacement-sensitive device. As shown, a single rounded foot may be used, or a swivelling pad (Figure 9.12(b)). The quantity measured is the vertical movement of the stylus relative to the skid or shoe as it slides over the crests of the workpiece. T indicates a transducer.

The diamond stylus may have the form of a 90° cone or four-sided pyramid, with a standardized equivalent tip radius 2 μm and 10 μm and operative forces of 0.7 mN or 16 mN respectively. On specialized instruments on the finest textures, e.g. gauges, values of radius down to 0.1 μm and of operative forces down to 0.001 gf can be used. Electric transducers, followed by their attendant amplifiers are either displacement sensitive (independent of time) or motion sensitive (dependent on stylus

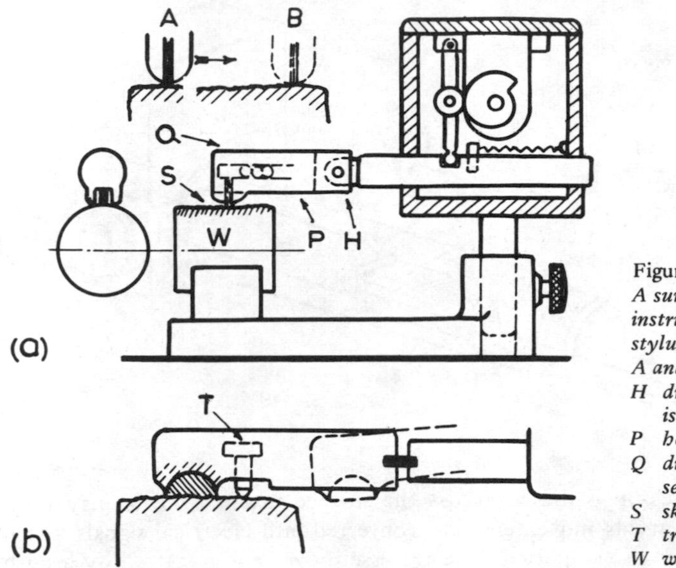

Figure 9.12.
A surface measuring instrument with a stylus:
A and B ends
H driving mechanism
P housing
Q displacement-sensitive device
S skids
T transducer
W work specimen

velocity). As examples, photo-electric and modulated-carrier transducers are displacement sensitive, while moving-coil transducers are motion sensitive.

Figure 9.13 shows a perspective view of the optical transducer which is incorporated in the stylus head of the Talysurf 10 surface-measuring instrument (Rank, Taylor-Hobson Ltd). Unbalanced signals from two photo-cells P are used to drive the R_a meter and the chart recording pen. From the lamp house assembly with its window A the beam is transmitted along a light guide G to a beam splitter B which incorporates the two photo-electric cells. In the 'zero' condition each cell receives the same amount of light from the beam splitter. The stylus bar is pivoted on a ligament hinge L, and at the end remote from the stylus S it carries a thin plate, or flag F, in which there is a narrow horizontal aperture. The plate is free to move vertically.

Movements of the stylus during its traverse are reproduced in a reverse sense by the flag which rises and falls in the space between the light guide and beam splitter. As a result, light passing through the aperture is no longer evenly distributed between the photo-cells, and, as the intensity in one rises, the other falls. Variations in the output are detected and amplified, and these are applied to drive the meter and the recording pen.

While many instruments are for laboratory use, the Surtronic 3 is a small compact instrument from the Taylor-Hobson range. Figure 9.14 shows it to be ideal for workshop use, for it can be taken to the component or can be used on a bench as it is battery driven. It has a miniature positive sensitive pick-up system, for profile integration is performed

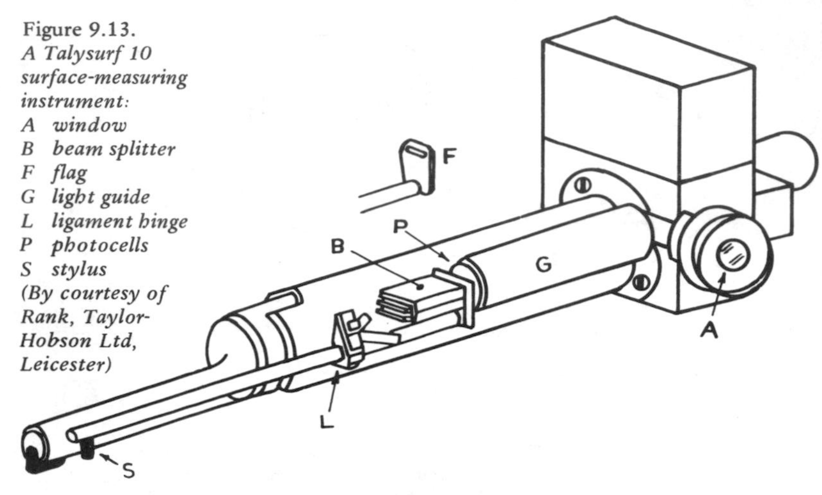

Figure 9.13.
A Talysurf 10 surface-measuring instrument:
A window
B beam splitter
F flag
G light guide
L ligament hinge
P photocells
S stylus
(By courtesy of Rank, Taylor-Hobson Ltd, Leicester)

while the pick-up is drawn across the surface towards the display unit. The vertical stylus movements are converted into electrical signals which are processed and amplified. The traverse movement is started by a push button, and, when completed, the R_a value is displayed. The single unit houses the 'start' button and the only two controls are range and cut-off. It also contains the liquid crystal display, traverse mechanism, pick-up holder, filter circuits and battery. The recorder has front loading for chart paper 50 mm wide, and push-button controls give a horizontal magnifications V_h of ×20 and ×100 and seven vertical magnifications V_v from ×100 to ×10,000.

Figure 9.14.
A portable Taylor–Hobson instrument Surtronic 3, battery driven

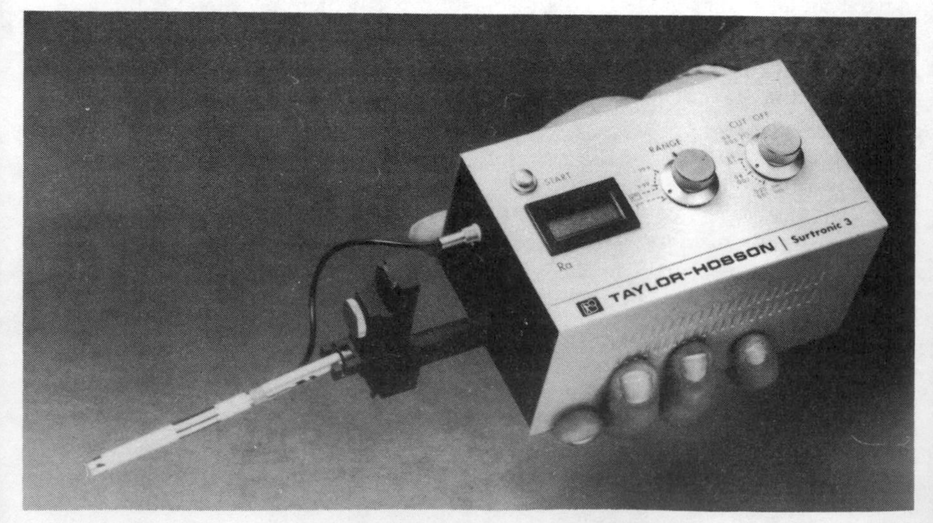

10

MACHINE SLIDEWAYS AND GUIDING SURFACES

LATHE AND OTHER MACHINE STRUCTURES

The stresses imposed upon a lathe bed are caused by two main forces.

(1) The downward pressure of the work acting on the tool.

(2) A pressure tending to force the tool away from the work in a horizontal direction.

There is a third force caused by the traverse of the tool in the longitudinal direction, but this has little effect, and the resultant force is largely one of torsion. This torsional stress tends to deflect the bed far more than the bending stresses, so that a box or tubular section is required. The problem is complicated by the necessity of providing large apertures for the escape of cuttings, a very necessary feature when using carbide tools.

Figure 10.1(a) shows the design of a bed (Dean, Smith & Grace Ltd) with the saddle in Figure 10.1(b). Vibration problems are eliminated by making the structure with diagonally disposed hollow bars at frequent intervals to shorten both vibrations and reactions, for the periodicity of vibration on any bed is higher; it is also staggered, and therefore less consequential, for the amplitudes of the waves is less. On the bed shown ease of swarf clearance is apparent, while the flat guideways are of hardened steel so that separate tailstock guides are not necessary. Another feature is that ease of travel of the saddle along the bed is assured, for the slideways are lined with low-friction plastic material which reduces stick-slip and wear and protects against abrasion. Automatic lubrication is provided, and the sensitivity of saddle movement is such that incremental movement of 0.0001 in (0.0025 mm) is obtained.

SLIDEWAYS

These form the basic elements of all types of machine tools, for they supply the mechanical guidance of tools and slides upon which the accuracy of the product depends. The shapes of the slideways are determined by a number of factors, including the load to be carried, the direction of the cutting forces and the position of the element which is used for transmission.

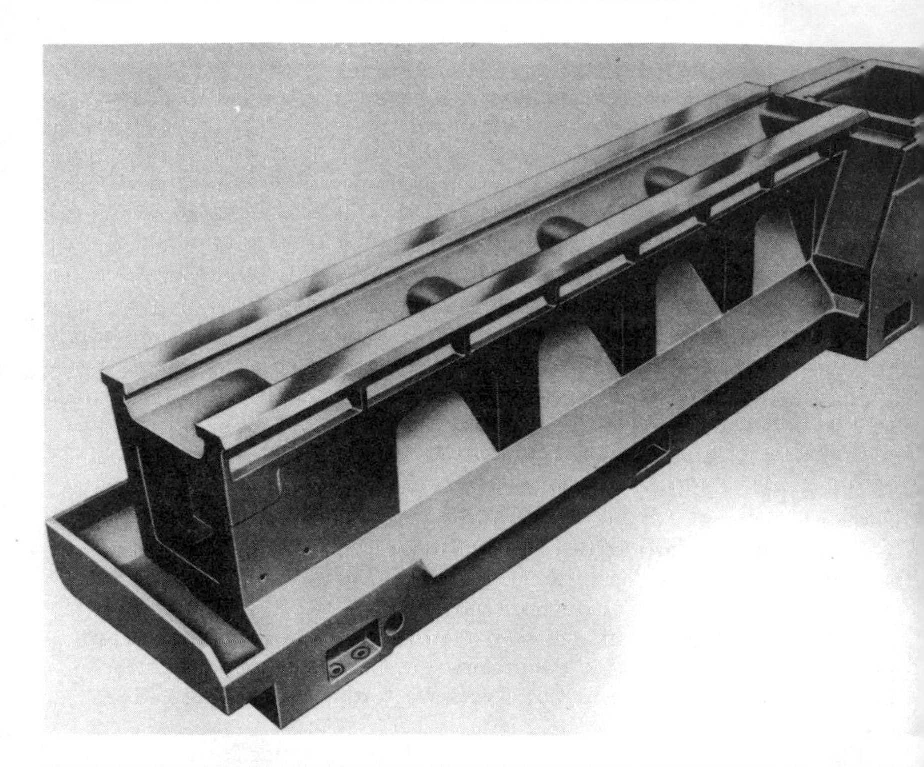

Figure 10.1.
Views of bed and saddle guideways on lathe
(By courtesy of Dean, Smith & Grace Ltd, Keighley)

For lathe beds both flat and vee guides are in evidence, the claims for the latter being automatic adjustment, for gravity acts as a closing force which keeps the surfaces in contact. Cross winding which causes the saddle to tend to wedge is prevented, and side strips are not required. The saddle is unaffected by wear, and metal particles do not easily adhere to an inverted vee.

The drawback is the lack of bearing surface which can result in rapid wear. Also the inverted vees weaken the saddle which has a long unsupported span across the bed. This cannot be strengthened without weakening the slides of the compound rest. In contrast the flat top bed has an abundance of bearing surface with the saddle well supported so that accuracy is maintained for a long period.

NARROW GUIDE PRINCIPLE

To reduce cross winding when traversing the saddle along the bed, the narrow guide principle is adopted. Flat top beds guide the saddle by the side shears, retaining plates being fitted under the saddle to prevent any lifting tendency. A taper strip on the side shear is used to adjust the saddle if wear takes place. The narrow guide restricts the guiding distance between the shears to a short dimension AB as shown in Figure 10.2(a) instead of the saddle spanning the bed, i.e. clearance is left at C. There is a diminution of power required to traverse the saddle along the bed by reason of the reduction of the frictional resistance caused by the pressure against the guiding edges, but it is important to realize that the success of the design is entirely dependent upon the position of the traversing members, rack and lead screw, which are in close proximity to the guide shown.

Figure 10.2(b) shows a section of the bed for a Harrison lathe with flat and inverted vee guideways; one of each type is used for location of the saddle and tailstock respectively. The object of this separate location is that the height of centre of the tailstock is unaffected by any wear of the bed caused by the saddle traverse, which is mainly near the chuck, if any occurs. However, to prevent wear the bed is milled and then slideway ground; it is then induction hardened and finally lightly ground to complete the operation. The materials used in the casting are 3.3% carbon, 0.2% phosphorus, 1.75% silicon and 0.5% manganese, with trace elements of nickel, chromium, molybdenum, copper, vanadium, aluminium and tin.

The method of swarf disposal is shown and consists of employing diagonal ribbing down which the cuttings fall onto a wheeled tray for easy removal. On larger lathes it is often necessary to install a conveyor of the drag type behind the machine, so that removal of the cuttings is continuous with the running of the lathe. Some complication may take

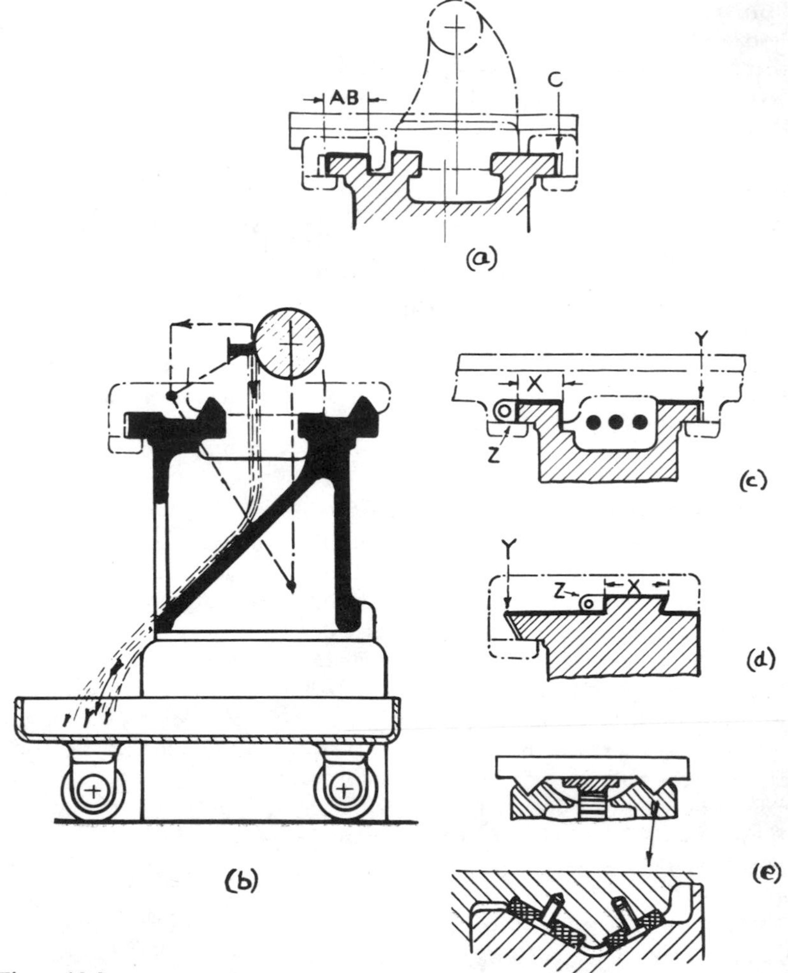

Figure 10.2.
Types of guideways used on a range of machine tools

place when a copious supply of suds is used; the arrangement is then for the suds to run into a trough and thence back to the tank.

Another flat top bed design is shown in Figure 10.2(c), which is for a horizontal boring machine. This is more simple than those previously described, for no separate guide for units such as tailstock is required, the location being for the table only. The narrow guide is shown at X, Z being the taper-adjusting strip, and Y indicates the clearance. Another design for a milling machine is given in Figure 10.2(d), again with X denoting the narrow guide, Z indicating the taper strip, and Y the clearance.

PLANING MACHINES

Both flat and vee guides are used; an advantage claimed for vee guides is that they are self-compensating in the vertical plane, but a greater advantage is that they are self-compensating for wear in the horizontal plane. With correctly designed vee slides it is virtually impossible to make a planer table move up the incline under normal cutting conditions. The reason for the use of flat guides is that they are easier and cheaper to manufacture, particularly on large machines where the fitting and alignment of four vee faces is not an easy proposition. Friction is less on flat than on vee guides, and in order to overcome this drawback, non-metallic table ways are used (Figure 10.2(e)).

As shown, plastic plates of laminated phenolic plastic material are secured to the table by phenolic plastic pins. It is claimed that table-bearing loads of 100% greater than formerly possible with cast iron slides can be employed. Also the low coefficient of friction and good heat-insulating properties of the plastic permit increased cutting and return speeds. Cutting and scoring of the bed is eliminated as the plates have no welding properties.

CYLINDRICAL GUIDES

This form of guide has wide application to the pillars of radial drilling machines; the advantage is that a circular guide facilitates rotation and elevation of the arm, while clamping devices are simple and efficient. The single round guide must support not only the weight of a cantilever arm but also the saddle travelling along the arm. The problem is to prevent the sleeve on the column from wedging and giving the condition of 'cross winding'. On a radial drilling machine this means providing a long bearing sleeve on the column. In all cases of the narrow guide the success of the design is dependent upon the traversing member, rack, screw or piston, which must be in close proximity to the guide, i.e. for a radial drill as close as possible to the sleeve. From Figure 10.3 the length L can be

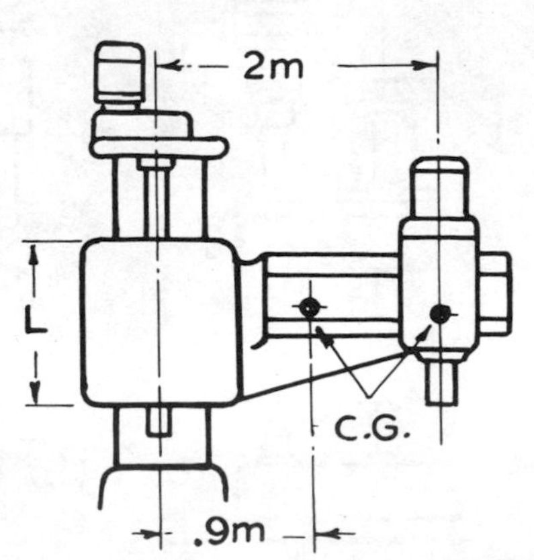

Figure 10.3. *A cylindrical guideway for the arm of a radial drilling machine*

calculated to ensure free movement of the arm on the column when operated by the screw. If the weight of the arm is 1000 kgf, the weight of the saddle 250 kgf, with μ=0.3 between the column and arm, then if the arm is considered on the point of wedging, then the upward force must equal the downward force.

Therefore 1,250 = $2F$, where F is the frictional resistance. F = 625 kg; thus clockwise moments equal anticlockwise moments or

$$(0.9 \times 1{,}000) + (250 \times 2) = N \times L$$

$$1{,}400 = N \times L$$

Therefore

$$N = 1{,}400/L$$

$$625L/1{,}400 = 0.3$$

or

$$L = 0.672 \text{ m}$$

Hence L must be greater than 67.2 cm if sliding is to take place.

DESIGN FEATURES

Tube has a range of applications with advantages inherent in the tubular cross section. For a given weight, a tubular member provides the optimum in material distribution and strength under normal loading conditions. Structures built of such sections tend to absorb and localize shock caused by impact, thus minimizing damage. Under complex imposed stresses a tube most nearly meets the conditions of an ideal member, while under dynamic loading tubing shows greater rigidity, with a higher frequency and smaller amplitude of vibration than any other section, including the solid round in which the inner portion does little more than increase the weight.

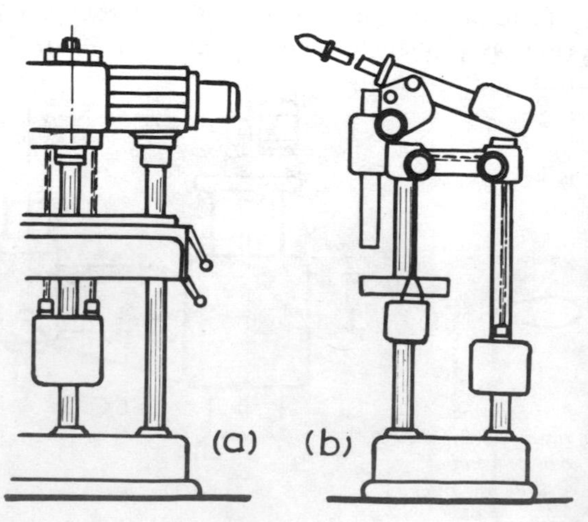

Figure 10.4.
The use of bright steel bar to simplify machine construction

With the idea of saving cost of patterns for machine bodies and to facilitate construction without the machining of slideways and fitting of slides, the writer has designed and constructed in the workshops of the Keighley Technical College, the two machines shown in Figure 10.4 (a) and (b); these are a keyway cutting machine and a mandrel press respectively. Both employ bright steel bars to form the main supports, the three bars being connected at the top and bottom to produce the same result as an expensive body casting but with considerable economy. The only machining of the bars is turning at both ends to fit brackets at the top and the base plate at the bottom. The rear member in each case has the dual purpose of a support and to act as guide for the balance weight for the table. Clamping of the table on the bars is by the pad and socket method by using the ball-ended levers.

ANTI-FRICTION SLIDEWAYS FOR MACHINE TOOLS

These slideways provide the following advantages.

(1) Increased sensitivity of precision traverses.

(2) The possibility of increased speed of traverse with increased life over conventional slides.

(3) Easier manual traverses.

(4) Accuracy with regard to positioning a machine member in a given location.

(5) Reduction in the driving power required and in heat generation.

Good results in overcoming the phenomenon known as slip-stick, which prevents easy starting from rest, have been obtained by the use of ball or roller designs. Ball-bearing guideways which are used to substitute rolling for sliding friction, are suitable for the lighter type of machine tools as for the table traverse of a cutter grinding machine of Figure 10.5. Ease of movement is obtained by locating the table on one ball track with the right-hand side which has only a single flat surface in contact with the balls. Hardened and ground inserts fitted into table and bed units form the guideways for the balls which are carried in chain cages. The hardness of the inserts should be the same as that of the bearings, i.e. 64 Rockwell C in order to resist Brinelling. In some designs the table is kept in contact with the ball slides by a heavy coil spring which can be

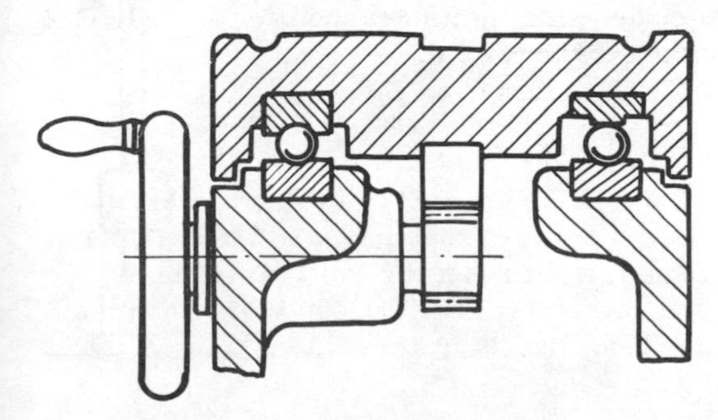

Figure 10.5. *Ball-bearing guideways for a cutter grinding machine*

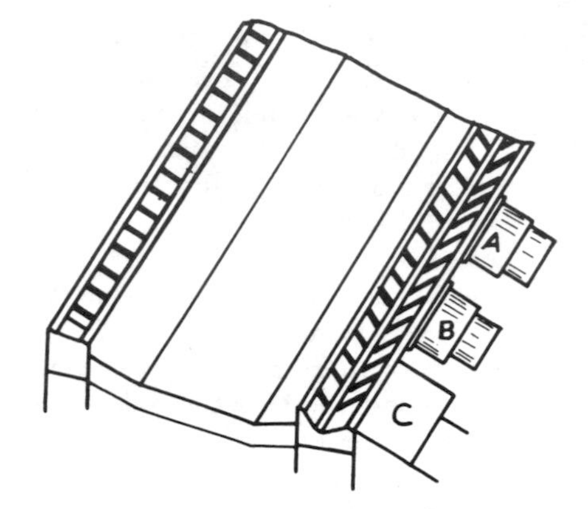

Figure 10.6.
Roller bearing guideways for a jig boring machine:
A and B Selsyns
C cover

adjusted so that the table movement is unimpeded by sliding friction.

ROLLER BEARING DESIGN

Some precision machine tools such as grinding machines and jig boring machines are located for the table traverse on one vee and one flat guideway. This is shown in Figure 10.6 which illustrates the bed of a jig boring machine with the roller bearings in position. The object of the anti-friction location on this machine is that the accuracy of co-ordinate setting, i.e. the movement of the x and y axes of the table, require to be 0.0001 in (0.0254 mm).

After co-ordinate positions have been pre-selected by rotary controls, a button is depressed, and the table automatically traverses in the x and y axes to bring it very close to the final position. Measurement during this part of the cycle is automatic and is effected by means of racks and coarse- and fine-reading Selsyns. When the table comes to rest, the operator completes the final positioning by hand and by using optically viewed scales. The Selsyns are shown at A and B and are associated with the longitudinal movement of the table.

A similar arrangement is employed for the transverse movement of the saddle, but in this case there is a central roller vee guide, and two widely spaced flat roller guides which provide support for the ends. One of the flat roller guides is under the cover C.

CROSSED ROLLER CHAINS

Figure 10.7 shows a system (Skefko Co. Ltd) to give increased mobility to machine slideways. Here a series of rollers, held in retainers, forms a flexible chain; the assembly is such that the axis of rotation of adjacent rollers are at 90° to each other. The diagram shows an arrangement for a machine tool slide, the chains being fitted between tracks

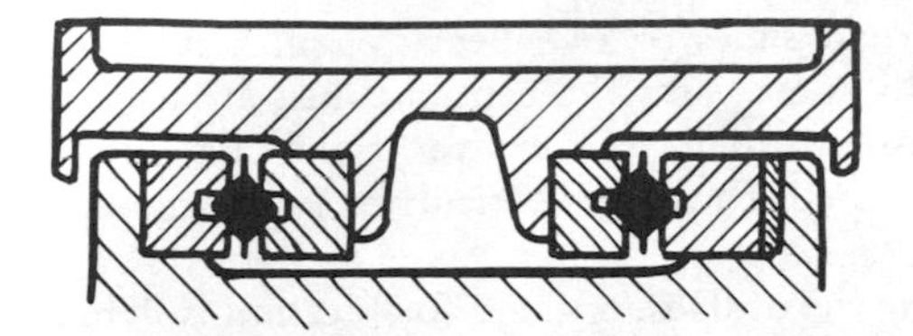

Figure 10.7.
A crossed roller chain for a machine tool slideway (By courtesy of Skefko Co Ltd)

with vee-shaped grooves. They can be mounted so that the retainers are in either the vertical or the horizontal plane as required and the slide bearings can be pre-loaded by using a slightly tapered track and adjustable gib.

The carrying capacity depends upon track hardness, e.g. the maximum permissible load per roller, with a track hardness of 300 Brinell, ranges from 32 lb (15 kg) to 770 lb (349 kg) for the smallest and largest size of chain, and from 240 lb (109 kg) to 5500 lb (2,500 kg) for hardness values of 650 Brinell. The total load permissible is the carrying capacity per roller multiplied by the number of rollers supporting the load at any given time. The values relate to loads normal to the cylindrical surfaces of the rollers, and, to obtain the permissible vertical load, they must be multiplied by sin 45° (=0.71). This type of bearing can be used for circular tracks, as, for example, those used for large spindle bearings of boring and turning mills, but these are not recirculating.

BALL AND NEEDLE ROLLER BUSHINGS

Examples have been given where a solid round bar has been used for vertical guideways, but designs are available where vertical guideways are used for large and heavy machine tools such as horizontal boring and facing machines. Such guideways have traditionally been used for large presses for many years with advantageous features, some of these machines standing as high as a four-storey building. To reduce friction between the circular guides and the sliding member, ball bushings are available as standard units or, on very large machines, can be built into the design. Similarly, as an alternative, needle rollers can be used, in which case the movement of the needles is a compromise between rolling and sliding; the main advantage over ordinary roller bearings is that they can be used where space is too limited for any other type of bearing.

CENTRAL THRUST LOCATION

This type of construction for the location of a machine saddle can be an advantage, since on boring machines of more usual construction the pressure of the cut is taken on a side strip only. With a central thrust saddle location the backward pressure during machining is taken directly on two broad slides or two heavy circular guideways plus a rear flat bearing. In some cases the central thrust arrangement means that the saddle passes through the centre of the column, and while ample rigidity is obtained it may be difficult to locate the operating controls.

The advantages of using vertical round bars for location is demonstrated in Figure 10.8 which shows the Scharmann boring machine where saddle location is by two heavy front pillars; the spindle passes between them to obtain the advantages of a central thrust construction. The pillars are hardened and ground, while two mechanically adjustable bushes guide the headstock on each column. Clamping is effected hydraulically, independently of the guiding bushes. The short rigid spindle saddle is supported endwise against a rear upright; this unit and the guide bars are tied together by means of a box-section member at the top of the machine.

When round-bar guides move from vertical to horizontal positions, they are generally classified as beams and, in doing so, tend to be less desirable for guideways than their vee or square slide counterpart is. Nevertheless, there are exceptions, one of which is the massive round guideways used on some large extrusion presses.

Figure 10.8.
Vertical cylindrical guideways on boring machine
(By courtesy of Scharmann Machine Ltd, Solihull)

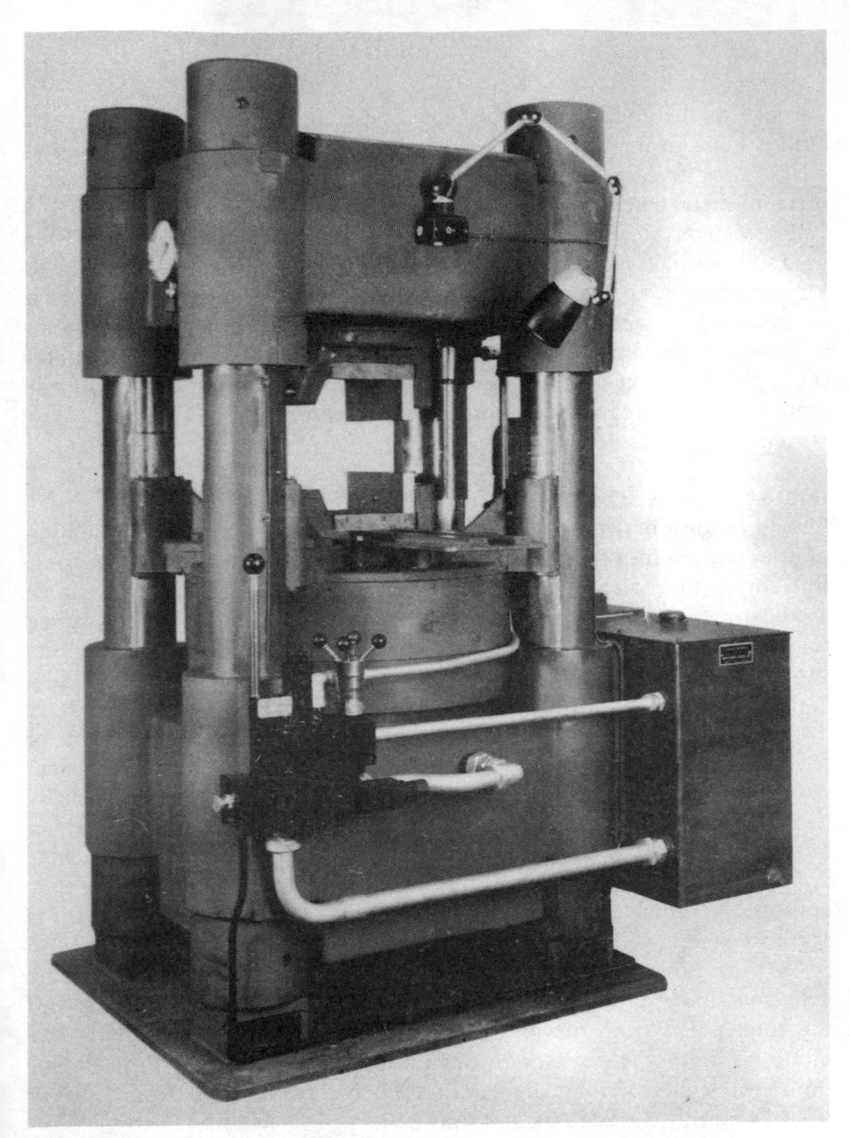

Figure 10.9.
A hydraulic press with cylindrical guideways
(By courtesy of Armstrongs (ENG) Ltd, Leeds)

When a number of round bars are used as columns, not only is there a saving in material and machining costs, but they have the advantage over other slides in lightness. Where these guides are of tubular construction they will absorb and localize shock, for, under dynamic loading, tubes show greater rigidity (with a higher frequency and smaller amplitude of vibration) than any other section. One example of using tubes horizontally is when in a milling machine the tubes are fastened to the underside of the table and slide in semi-circular bed ways. Drilled holes in the tubes move past a pressure oil inlet in the bed, so that oil escaping from other holes causes flotation of the table and prevents metal-to-metal contact. Surface wear is thus negligible.

Presses constitute a large field in which cylindrical guideways have been used. Whilst deformation must not allow the parallelism of the ram to vary in relation to the platen, elongation is not detrimental as long as it symmetrical on steel structures. An excellent design of a 1,000 ton press is shown in Figure 10.9. (Armstrongs (Engineers) Ltd) and illustrates the rigidity obtainable by the four columns. The compactness of the mounting of the hydraulic equipment is also apparent.

HYDRO-STATIC BEARINGS

Sliding friction between two surfaces can be reduced by the introduction of a supporting film of fluid. In the commonly used sliding surface bearing, the lubricant is merely delivered to a point from which it is carried between the bearing surfaces by their relative motion. A minimum speed is required to ensure that the oil will separate the parts and will provide full film support. Under these conditions the bearing is operating hydro-dynamically.

In contrast, hydro-static bearings, as indicated in Figure 10.10 are designed to operate with the load always fully supported on a film of fluid. The lubricant (or air or pressurized gas) is supplied under pressure to one or more recesses in one of the mating bearing members. Usually these recesses are located in the stationary member or pad. The other bearing member is usually termed the runner but is, in practice, a machine table or saddle. The surfaces may be of any practical shape so long as they conform in contour with each other. For most hydro-static bearings there are several pads supporting each other.

In operation the oil supply must be started and the fluid must be delivered under pressure to the pad recesses. Pressure is increased until the runner lifts from the pads and is fully supported by the fluid. The oil flows from the recesses through the separation areas, the escaping fluid being collected for recirculation. Film thicknesses range from 0.001 in (0.0254 mm) to 0.010 in (0.254 mm).

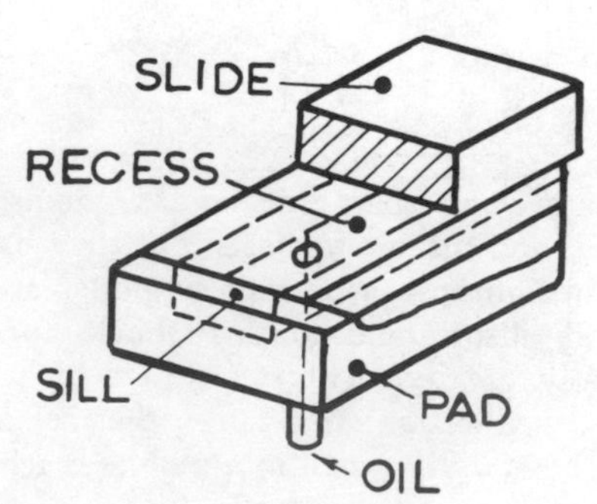

Figure 10.10.
Details of the construction of a hydrostatic bearing

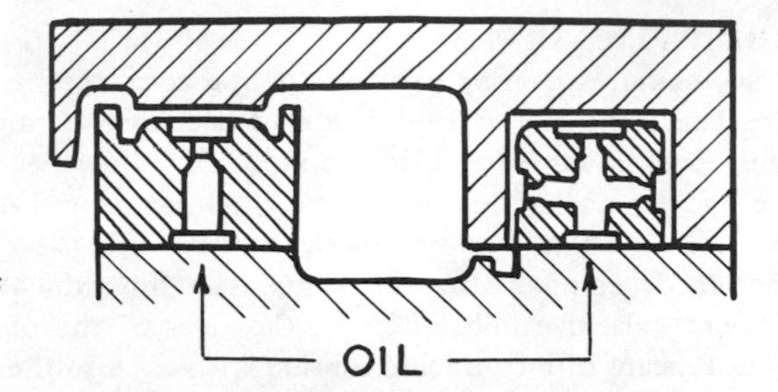

Figure 10.11.
Hydrostatic bearings applied to the table of a grinding machine

An advantage is that a hydro-static bearing can support its full load at very slow speeds or with no movement at all. Friction in the bearing is due only to the viscous drag of the oil film and is proportional to the relative velocity between the runner and pads. Thus heavy loads can be supported and set in motion with no starting friction (or stiction) and can then be moved at slow speeds with little effort.

The features given renders hydro-static bearings suitable for NC machine tools. The stick-slip effects encountered with slides and surfaces can be eliminated by their use. They ensure low bearing friction with long service life, while machine members do not need to be scraped to a finish required for normal slideways. If required operation speeds can be high, cylindrical externally pressurized liquid bearings have been designed to run at 50,000 rev/min, while gas bearings can operate at even higher speeds.

Figure 10.11 shows hydro-static bearings applied to the sliding table of a cylindrical grinding machine where good control was required in a horizontal direction with less critical vertical control. The coefficient of friction is very low, about 0.0001 or less, and the error in the motion of the moving member, in relation to that of the stationary member, contains an error usually less than one-tenth of the individual errors in the elements from which the bearing is constructed. A slide, for example, made from several elements each with individual tolerances of 0.0001 in (0.002 mm) will move along a line which is straight to within ten millionths of an inch.

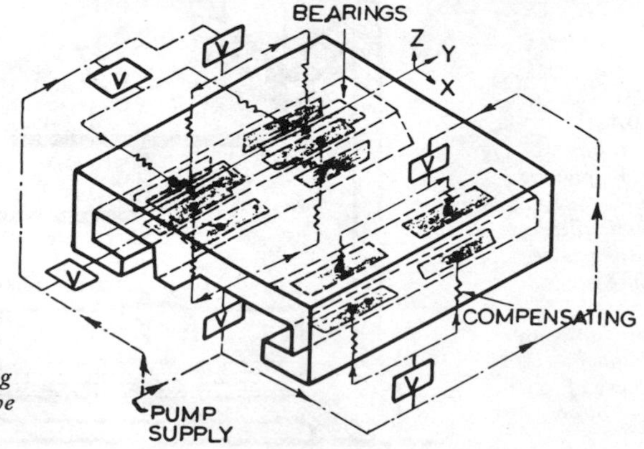

Figure 10.12.
A twelve-bearing design for a lathe saddle

LATHE APPLICATION

To solve complex bearing problems digital computers are being employed. Figure 10.12 shows the application of a twelve-bearing lathe saddle, the bearings being used to restrain the saddle in all degrees of freedom except in the direction of translation in the axial direction, the y axis. Lubricant is fed from a single pump to each of the bearings by way of individual restrictors or compensating elements. The function of the restrictors is to prevent excessive imbalances in the flow of the bearings and also to provide a means of introducing oil film stiffness into the system.

For the lathe saddle, relative motion is along the y axis, with possible cutting loads of the x, y and z directions, applied at the tool point. Thus there are twelve reaction points to any cutting load. The circuit for a lathe is complicated in that motions in one direction can produce reactions and moments in other directions, which in turn depend on the supply circuitry and pre-load pressures, and all of these can vary. For most machine tool applications, tool loads are the given quantities and tool movements the desired outputs.

PLANING MACHINE BEDWAY

While not strictly hydro-static operation, a method of flotation for the table of a planing machine (Waldrich-Coburg) is shown in Figure 10.13. Oil is supplied by an electric pump D through the magnetic filter A. The front faces of the guideways are provided with end pieces which mesh into the longitudinal slots of the table slideways. This provides an enclosed pressure oil channel K below the table guide to ensure an even rise and distribution of pressure oil for the lubrication grooves spaced throughout the entire length of the guides. The lubrication slots, arranged at a

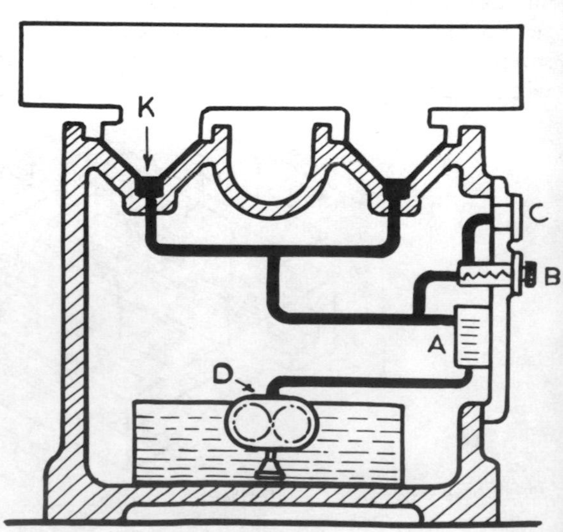

Figure 10.13.
An oil flotation system for a planing machine:
A magnetic filter
B regulating valve
C pressure gauge
D pump
K enclosed pressure oil channel
(By courtesy of Waldrich–Coburg, Germany)

close distance, have vee-shaped scraped lubrication pockets of such a form that a hydro-dynamic lubrication process is provided during the table movement. This system provides an adequate oil film which reduces friction to a minimum. The oil viscosity is 290 SUV at 100°F.

In addition, a regulating valve B is provided to adapt the oil pressure to the table load. The pressure range of 0.1 to 0.5 atmospheres can be controlled and checked by the pressure gauge C. An important feature is that the pump D is electrically interlocked with the main driving motor of the machine. Thus in the event of failure of the lubrication the main motor is immediately and automatically switched off. While the table guideways are provided with synthetic material, this is only to prevent scratching or scoring of the slideways by foreign matter which may find entrance between the members. The reduction of friction is taken care of by the pressure oil system, which maintains an unbroken oil film.

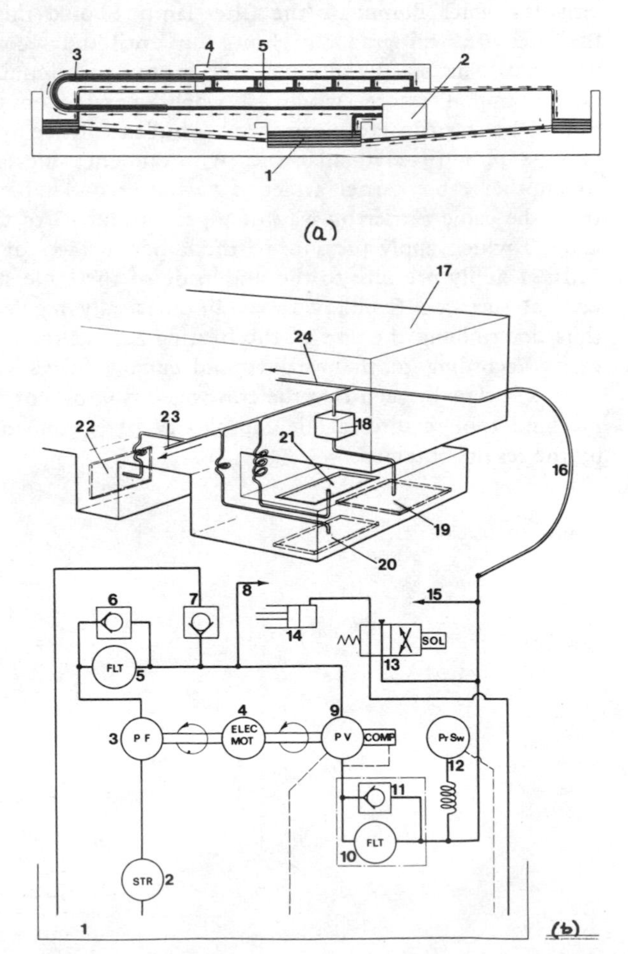

Figure 10.14.
A hydrostatic table lubrication of a plano-milling machine
(By courtesy of Butler Machine Tool Co. Ltd, Hailfax)

The hydro-static table lubrication for a Butler plano-milling machine is shown in Figure 10.14(a). Oil is pumped from the bed sump 1 into the hydraulic unit 2 and then, under pressure, through cable carrier 3 to table 4 and hydro-static pads 5. Oil then returns by gravity to either the bed centre or the end collecting troughs and sump.

Figure 10.14(b) shows an electric motor 4 that drives the pump 3 which draws oil through the strainer 2. Oil then passes to a variable-delivery pump 9 by way of filter 5 which, if the oil becomes dirty, is by-passed by the check valve 6. The fixed-delivery pump 3 always delivers more oil than required by the variable-delivery pump, and the excess is returned to the sump by way of the excess boost check valve 7.

Oil from the variable-delivery pump is passed through filter 10 which contains an integrally-mounted check valve 11. Should filter 10 become dirty, the pressure rise in the check valve operates the first of two micro-switches which illuminate the filter lamp. Should this lamp be ignored, the pressure will gradually build up until the second micro-switch is operated. This stops the table feed motor and illuminates the pressure-failed lamp. Pressure switch 12 when operated ensures that the table cannot be started before pressure has built up in the hydro-static bearings.

Pipe 15 is fitted to machines with tandem tables and carries pressure to another cable carrier attached to the second table. Pressure is taken from the cable carrier by way of pipes 24 and 23 to the bearing pads 19 and 20 which apply pressure to the upper surfaces of the bed slideways. Pads 21 apply pressure to the underside of the table strips and 22 to the vertical surfaces. Capillary tubes automatically regulate the flow of oil, thus determining the size of the bearing gap, as the load on the bearings varies according to the weights and cutting forces acting on the table.

It will thus be seen that the controlled flotation of the table of a heavy machine tool requires a fair complexity of equipment but is warranted by the results obtained.

11

LABORATORY EXPERIMENTS AND WORKED EXAMPLES OF MACHINING PROJECTS

Unless a narrow field of investigation is considered, a programme of metal-cutting experiments can reach a large dimension owing to the many variables which have to be considered. Principally they are as follows.

(1) Tool angles.
(2) Tool material.
(3) Workpiece material.
(4) Variations of cutting speed, feed, depth of cut.

Each of these has many subdivisions so that certain limitations must be established for any given test.

Any motor-driven machine tool lends itself to an analysis of power requirements for a given operation by the connection of electrical recording instruments, whilst dynamometers to measure cutting forces in two or three directions are available. These are based upon spring deflection, hydraulic or pneumatic pressure, or even electrical load cells, a dial or other display giving a visual reading of the force measured. The machine tools most frequently used for such experimental work are lathes, drilling, shaping and milling machines. First we shall consider a lathe application.

EXPERIMENT WITH CUTTING TOOLS AND LATHE DYNAMOMETER

OBJECT

The object was to investigate the cutting forces during rough turning with a variable cutting speed.

EQUIPMENT

The equipment used was a Dean Smith and Grace 8½ in centre lathe fitted with a dynamometer to measure cutting forces in three directions (Figure 11.1).

(1) Tangential force.
(2) End force.
(3) Feed or side force.

Each of these forces is shown as a deflection of a calibrated spring in millimetres by means of dial indicators.

Figure 11.1.
A lathe fitted with a dynamometer recording three force directions

TEST MATERIAL

The test material was a mild steel bar 76 mm in diameter.

METHOD

A series of cutting speeds was selected with a constant depth of cut (3 mm) and a constant feed rate of 0.38 mm/rev of the spindle. Cuts were taken, and the results shown in Table 11.1 were obtained.

Table 11.1.

Spindle speed (rev/min)	Cutting speed (m/min)	Depth of cut (mm)	Feed (mm/rev)	Tangential deflection (mm)	End deflection (mm)	Side or feed deflection (mm)
25	6.1	3	0.38	0.1270	0.0406	0.1372
37	8.8	3	0.38	0.1346	0.0432	0.1397
59	14.3	3	0.38	0.1346	0.0483	0.1397
88	21.0	3	0.38	0.1372	0.0492	0.1321
140	33.5	3	0.38	0.1524	0.0508	0.1016
208	50.0	3	0.38	0.1524	0.0559	0.0762
280	66.8	3	0.38	0.1651	0.1270	0.0762

The conversion factors required to change the deflection readings to kilograms are given as a force divided by a deflection equalling a constant and are for the tangential force 2289, for the back or end force 887 and for the feed or side force 1,290; the deflections in millimetres are multiplied by these conversion factors to give the forces in kilograms. These actual forces are shown in Table 11.2.

Table 11.2.
Actual forces (kg)

Tangential	End	Side
291	36	177
308	39	180
308	43	180
315	44	170
349	45	131
349	50	98
378	114	98

$$\text{hp at the tool tip} = \frac{\text{vertical load} \times \text{cutting speed (m/min)}}{4573}$$

and therefore for

$$\text{cut 1:} \quad \text{hp} = \frac{291 \times 6.1}{4{,}573} = 0.388$$

$$\text{cut 2:} \quad \text{hp} = \frac{308 \times 8.8}{4{,}573} = 0.593$$

$$\text{cut 3:} \quad \text{hp} = \frac{308 \times 14.3}{4{,}573} = 0.963$$

$$\text{cut 4:} \quad \text{hp} = \frac{315 \times 21}{4{,}573} = 1.446$$

$$\text{cut 5:} \quad \text{hp} = \frac{349 \times 33.5}{4{,}573} = 2.556$$

$$\text{cut 6:} \quad \text{hp} = \frac{349 \times 50}{4{,}573} = 3.815$$

$$\text{cut 7:} \quad \text{hp} = \frac{378 \times 66.8}{4{,}573} = 5.521$$

It can be seen from the results obtained that as the cutting speed increases the tangential and end forces increase fairly uniformly but that the feed pressure or sideways force gradually decreases as the cutting speed increases. A graph plotted of cutting speed and horsepower shows the latter increases uniformly with the cutting speed.

For the purposes of these test cuts a cemented carbide tool with a 10° positive rake was used; it had a standard 11° front clearance.

EXPERIMENT TO INVESTIGATE THE RELATIONSHIP BETWEEN FEED, DEPTH OF CUT AND REQUIRED POWER

EQUIPMENT

The equipment was a Ward combination turret lathe equipped with a wattmeter instead of a dynamometer to monitor the main motor power input, a cemented carbide cutting tool with a 10° rake, top and side, and an 11° front clearance, and a revolving centre.

TEST MATERIAL

The test material was a mild steel bar 112 mm in diameter and 1,300 mm long.

METHOD

The bar was gripped in a self-centring chuck and supported by the revolving centre which was mounted in the turret. A cutting speed of 63 m/min was selected, corresponding to 177 rev/min, and a wattmeter reading was taken with the machine running light (without cutting) but with the feed engaged. The wattmeter corresponded to 1.09 hp which

Table 11.3.

Test no. (mm)	Feed (mm/rev)	Depth of cut (mm)	(cm^3/min)	Input (W)	Gross power (hp)	Net power (hp)	(hp/cm^3)
1	0.5	0.79	24.42	1,416	1.9	0.81	0.0329
2	0.5	1.57	48.83	2,040	2.73	1.64	0.0335
3	0.5	2.36	73.41	2,592	3.47	2.38	0.0323
4	0.5	3.18	98.65	3,048	4.08	3.99	0.0305
5	0.67	0.79	33.43	1,680	2.25	1.16	0.0347
6	0.67	1.57	67.02	2,232	2.99	1.90	0.0280
7	0.67	2.36	100.50	2,904	3.90	2.81	0.0280
8	0.67	3.18	135.00	3,600	4.82	3.73	0.0274
9	0.98	0.79	48.99	1,920	2.57	1.48	0.0299
10	0.98	1.57	97.83	2,832	3.80	2.71	0.0274
11	0.98	2.36	146.80	3,720	5.00	3.91	0.0268
12	0.98	3.18	196.60	4,560	6.11	5.02	0.0256
13	1.95	0.79	97.50	2,592	3.47	2.38	0.0244
14	1.95	1.57	195.60	4,176	5.66	4.57	0.0231
15	1.95	2.36	293.30	5,640	7.56	6.47	0.0219
16	1.95	3.18	393.20	7,200	9.65	8.56	0.0213

746 W = 1 hp.

was assumed to be a constant irrespective of whether the machine was cutting or not. As shown in Table 11.3, the feed and depth of cut were varied, and the various values were tabulated. It can be seen from the final results that the machine ran more economically as the feed was increased, and therefore, as feed is increased, less power is required to remove a given amount of material.

EXPERIMENT TO COMPARE CERAMIC AND CEMENTED CARBIDE TOOLS

OBJECT

The object was to compare the characteristics of the two types of tool material with respect to life at high speeds, work surface finish and power requirements.

EQUIPMENT

The equipment was a Ward combination turret lathe, a carbide tool with a 7° negative top rake and a 7° side clearance, a ceramic tool of the same geometry but with a chip breaker mounted and a wattmeter coupled to the 7½ hp motor.

TEST MATERIAL

The test material was a mild steel round bar, 112 mm in diameter.

METHOD

A range of suitable cutting speeds was selected, and test cuts were taken with each tool at each speed. The feed rate and depth of cut remained constant throughout the tests at 0.1 mm and 2.5 mm respectively. Table 11.4 gives the results. The no-load input to the machine was measured to be 1.1 hp.

Table 11.4.

Spindle speed (rev/min)	Cutting speed (m/min)	Input (W)	Net power (hp)	Tool
88	32	2,610	2.4	Carbide
88	32	2,760	2.6	Ceramic
140	50	3,135	3.1	Carbide
140	50	3,360	3.4	Ceramic
208	75	4,250	4.6	Carbide
208	75	4,625	5.1	Ceramic
336	120	6,120	7.1	Carbide
336	120	6,685	8.1	Ceramic
500	180	6,790	8.0	Carbide
500	180	7,760	9.3	Ceramic

These tests enabled three conclusions to be drawn.

(1) The claim for superiority of ceramic tools at high cutting speeds was substantiated, for the spindle speed was increased to 1,000 rev/min for a final cut with the ceramic tool only which gave a cutting speed of 350 m/min. The finish on the workpiece was further improved, whilst signs of deterioration of the tool were negligible; the cemented carbide too, however, without this final test showed obvious signs of use, and there was some wear on the cutting edge. (In this final test the wattmeter was off its calibrated scale, and a reading could not be obtained.)

(2) Against this, the claim that a ceramic tool requires less power than an equivalent cemented carbide tool was not substantiated, for, as Table 11.4 shows, a lower power input was used in every case for the carbide tool, even when it was showing signs of wear. It was thought that the chip breaker fitted to the ceramic tool was absorbing some power but probably not sufficient to account completely for the difference between the two tools in terms of power input.

(3) The net horsepower readings in Table 11.4 show that the machine was overloaded by the tests and that any further tests of this type should be performed with a depth of cut of no more than 1 mm which is in line with current finishing practice in industry.

EXPERIMENT TO DETERMINE THE ROUGH TURNING CAPABILITIES OF A LATHE

OBJECT

The object was to determine if the 35 hp motor fitted to the machine was capable of substantial roughing cuts on steel forgings.

EQUIPMENT

The equipment was a 20 in centre lathe. A cemented carbide cutting tool with a 8° positive top rake, a test motor for the lathe of 35 hp, an alloy steel bar 230 mm in diameter, to an imperial standard of 50 ton/in^2 and an ammeter for the main motor current.

METHOD

A series of four test cuts were made by using a constant cutting speed of 88 m/min (at a spindle speed of 123 rev/min) with the feed and depth of cut varied as shown in Table 11.5. From the ammeter readings the gross horsepower was calculated, and 7.5 hp was deducted from it to show the power that was being absorbed by the tool. (7.5 hp had been observed as the input to the machine when it was running light without cutting.) The results are shown in Table 11.5.

These results were confirmed by a further series of tests using different parameters.

Table 11.5.

Cut no.	Feed (mm/rev)	Depth (mm)	Metal removal rate of chips (kg/min)	Gross power (hp)	Net power (hp)	Net power per metal removal rate (hp min/kg)
1	1.40	6.4	6.59	35.0	27.5	4.17
2	0.80	12.7	6.59	35.5	28.0	4.24
3	0.53	19.0	6.59	35.5	28.0	4.24
4	0.40	25.4	6.59	36.0	28.5	4.32

It was noted from the tests that the machine performed without signs of distress at full power from the motor and with various depths of cut. As a metal removal rate of 6 kg/min was acceptable, an increase of 5 to 40 hp was recommended to enable the main motor to keep a reserve of power. It was also observed from Table 11.5 that the most efficient cuts in terms of power consumption were the shallower ones with a high feed rate, cut number one being the most efficient.

It should be remembered that the actual power taken from a machine tool main motor is composed of two elements, the power used in the actual cutting process and the power losses in the transmission system. For many lathes the power used in cutting can itself be subdivided into that needed to rotate the work and that needed to drive the feed mechanism to the tool. It will be clear that, say, an eighteen speed constant-mesh gearbox driven by an a.c. motor will absorb much more power from that motor than would a three-speed gearbox driven by a variable-speed d.c. motor. In consequence it can be said that a lathe with the latter type of drive will have a much greater proportion of its motor power available for cutting, and, if thus equipped, it will usually have a separate feed motor, thus further increasing the power available to drive the work.

In planning machining cycles for multi-tool lathes it is desirable to avoid peak metal removal rates, even for short periods, which are beyond the designed maximum for the machine in question. This is particularly true when such a machine has to run for long periods on the same workpieces. Whilst calculations can give an indication of the value of such peaks, it is always safer to be able to measure them. Peak-recording meters are available which can be permanently installed on a machine and which will record only the highest power input to the machine. These can be readily reset, say, at the start of a shift, and give a clear indication that all is well. If a detailed analysis of the power input to a machine is required, a recording power meter can be fitted which continuously records the power input to a machine on a slowly moving roll of paper. Peak power demands can again be readily seen; it is easy to establish if these peaks are reaching the same value on successive workpieces, and remedial action can be taken by, say, regrouping tools, if the peaks are beyond the design parameters of the machine or its motor.

Large machine tools, particularly those intended for short runs of work including 'one offs', frequently have an ammeter mounted to enable the operator to satisfy himself that each cut taken does not overload the machine.

EXPERIMENTAL WORK ON DRILLING OPERATIONS

To carry out research into drilling and similar metal-cutting operations various types of dynamometers are available; they are generally mechanical, electrical, pneumatic or hydraulic in their operation, but occasionally load cells with amplifiers are found in modern designs. The first type relies on the deflection of a calibrated spring which is measured with accuracy and is displayed on a dial indicator, whilst the others rely on an electrical meter or a pressure gauge. In the latter case the pressure gauge displays the pressure on a spring-loaded piston and may take the form of a column of liquid in a glass tube. Monitoring the cutting parameters of the machine concerned during, say, a drilling operation can enable drill thrust, torque, gross and net power to be established, and these factors can be usefully compared between different metals.

Figure 11.2 shows a set-up for a drilling analysis which uses Solex equipment; the dynamometer is arranged to carry a chuck for holding the workpieces to be drilled. In operation the axial thrust of the drill against the workpiece is recorded on the manometer, whilst an arm contacts a dynamometer to record the torque exerted by the drill; a chart recorder may be used to give a permanent record of the results obtained. The manometer is calibrated up to 1000 daN, whilst the torque dynamometer has a maximum reading of 10 daN. These maxima approximately correspond to readings expected from drilling mild steel with a 25 mm (1 in) drill and from using a feed of 0.4 mm/rev (0.0156 in/rev),

Figure 11.2.
A dynamometer for recording thrust and torque when drilling

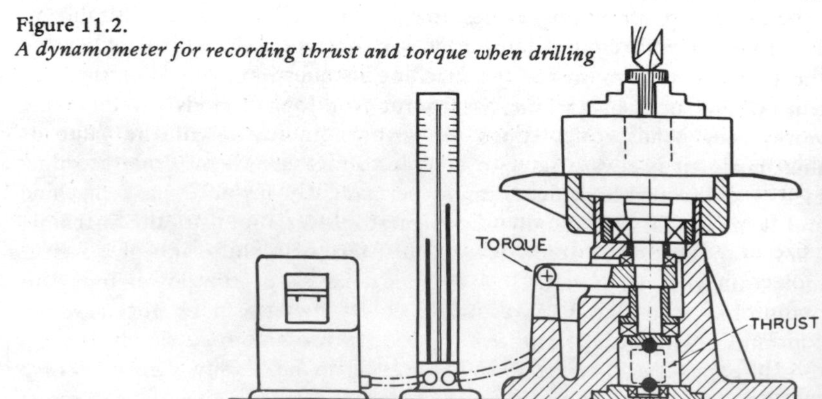

but a wide range of such instruments is available.

By using such equipment research may be undertaken to establish other factors such as the power per rate of drill penetration, the drill penetration rate per depth drilled between regrinding of the drill or the drilling efficiency expressed as the volume of metal removed per horsepower per minute.

For example to determine the torque T, thrust B or power for any combination of diameter and feed, the following equations have been determined by tests in which the drill diameter and feed were each varied separately. For chrome-vanadium steel

$$T = 1{,}840f^{0.78} \times d^{1.2}, \quad B = 53{,}000f^{0.78} \times d$$

and for cast iron

$$T = 380f^{0.6} \times d^2, \qquad B = 14{,}720f^{0.6} \times d$$

The total net horsepower developed at the drill point equals the horsepower due to the torque plus the horsepower due to the thrust which can be expressed as

$$\text{(total)} = 2\pi Tn/33{,}000 + Bfn/12 \times 33{,}000$$

In all the above equations, f is the feed (in/rev), d the diameter of drill (in) and n the rotational frequency of the drill (rev/min). To illustrate this point, a drill 1¼ in (32 mm) in diameter cutting at 60 ft/min (18 m/min) would need a torque of 110.8 lbf ft to keep it rotating whilst exerting a thrust on the workpiece of 2,430 lbf (1,093 kgf). If we substitute these values in the third equation, the horsepower due to torque is 3.69 and due to thrust is 0.016, a total of 3.71 hp or 2.8 kW. In most cases where the material that is being drilled does not work harden the thrust is only a small part of the total horsepower used in drilling but varies to some extent according to the type of thrust bearing if we consider the total input to the machine. It is therefore unimportant as far as provision of power is concerned but quite important in terms of the mechanical design of the machine.

Tables 11.6, 11.7 and 11.8 show the following data.

(1) The data obtained during cutting tests.

(2) The efficiency of the machine under different conditions.

(3) The net horsepower per volume of metal removed V (in^3) which shows the horsepower required per minute to remove 1 in^3 of the particular metal.

It will be seen that Table 11.7 shows that the efficiency falls as the size of drill is reduced, a conclusion which might be expected as the power losses due to friction, etc., remain more or less static whilst the power required to drive the drill is reduced for the smaller size.

Table 11.8 suggests that larger drills are more efficient than small ones, as the power required to remove a cubic inch of metal is smaller for larger drills.

Table 11.6.

Test piece	Drill diameter (in)	(rev/min)	Feed per revolution (in)	Torque (lbf ft)	Torque (hp)	Thrust (lbf ft)	Thrust (hp)	Total power at drill (hp)
Steel	0.5	444	0.009	14	1.18	725	0.007	1.19
	1.0	228	0.013	62	2.71	1,862	0.013	2.72
Cast iron	0.5	446	0.009	6	0.535	530	0.006	0.54
	1.0	229	0.013	28	1.22	1,088	0.008	1.23

0.5 in = 12.7 mm; 1 in = 25.4 mm; 0.009 in = 0.22 mm; 0.013 in = 0.32 mm.

Table 11.7.

Test piece	Drill diameter (in)	Gross power (kW)	Gross power (hp)	Power at drill (hp)	Efficiency %
Steel	0.5	1.6	2.14	1.19	55
	1.0	3.2	4.29	2.72	63
Cast iron	0.5	1.1	1.47	0.54	37
	1.0	1.72	2.30	1.23	53

Conditions as for Table 11.6.

Table 11.8.

Test piece	Power at drill (hp)	V (in^3)	Power per volume of metal removed (hp/in^3)
Steel	1.19	0.25	4.76
	2.72	0.75	3.63
Cast iron	0.54	0.25	2.16
	1.23	0.75	1.64

Conditions as for Table 11.6.

FURTHER EXPERIMENTAL WORK

Not all machine tools are so readily adaptable to such cutting tests, but the toolbox of a shaping machine may be adapted to take a load cell to measure the thrust upon the ram, and a power meter may be inserted in the motor power line to obtain further data. Equally, a load cell may be placed in the vise of a milling machine and light running and underload running readings taken from a wattmeter in the motor circuit. Variations in depth of cut, feed and rotational frequency of the cutter can be plotted, and the results can be analysed, whilst a vibrograph can be mounted on the table to study any vibrations which may be set up during cutting. A series of projects follows, mostly concerning plain milling, the more complex operations of universal milling being found in Volume 2 of this series.

EXAMPLE 1: MILLING A SPUR GEAR

Let us first consider milling a spur gear with, say, 35 teeth and 10 diametral pitch (Figure 11.3).

For gear-cutting projects it is often advantageous to make the blank of some soft material such as aluminium or plastic. This enables high-speed cutting to be used and saves considerable time so that more members of a class can take part in the exercises. Also, the wear on the cutter is reduced, and moreover the blank can easily be turned to a smaller diameter so that the effect of undercutting can be shown.

(1) The blank is mounted between the centres of the dividing head and tailstock and is connected by a driver to the plate of the head.

(2) A form cutter to suit the number of teeth and pitch is selected from the standard set and is mounted on the arbor.

(3) The cutter is then set central over the blank, either by measuring from a square set against the mandrel on each side or by letting the revolving cutter just touch the top of the blank to make a small flat and by setting the cutter in the centre of the flat.

(4) The next step is to set the sectors on the dividing head to suit the number of teeth required, i.e. $40/35 = 1\frac{1}{7}$ turns for indexing each tooth.

(5) The full depth of tooth is found by dividing 2.157 by the diametral pitch, i.e. 0.216 in, but, as two cuts will be required, the first cut should be about 0.180 in, leaving 0.036 for the finishing cut.

Figure 11.3.
A set-up for cutting gear teeth on milling machine

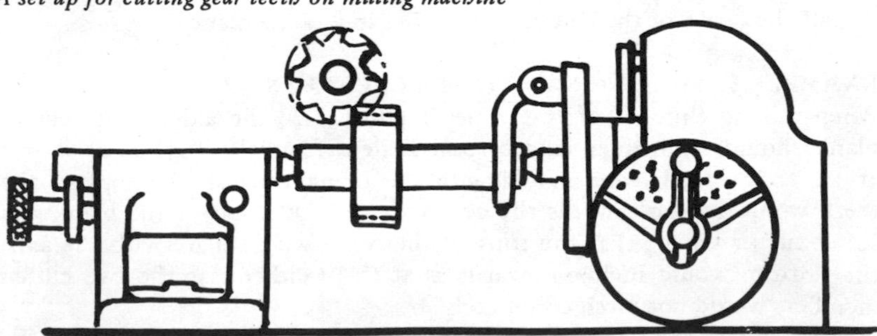

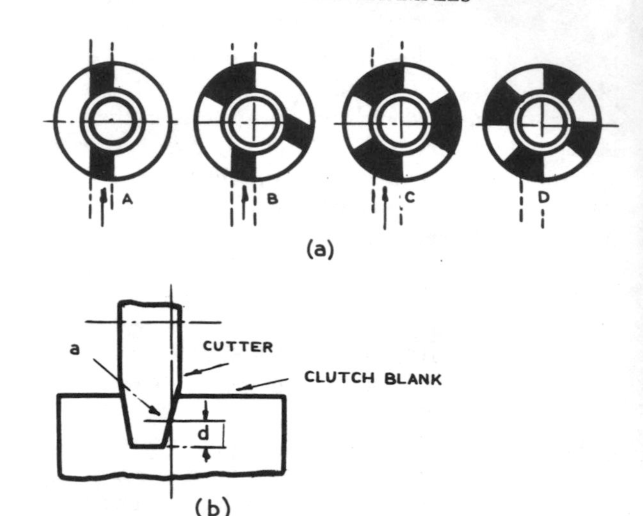

Figure 11.4.
A diagram showing the procedure in milling clutch teeth

(6) With stops set to give the correct table travel and with the cutter speed set at 100 ft/min, the feed can be engaged. After one tooth has been cut to the first depth and the cutter returned clear of the blank, the blank is indexed for each tooth in turn.

EXAMPLE 2: MILLING CLUTCH TEETH

In Figure 11.4(a), A, B and C show the first, second and third cuts for forming three straight teeth. The work can be held in the chuck of a dividing head set vertically. A plain milling cutter can be used, the side of the cutter being set to coincide with the centre line of the clutch. When the number of teeth is odd, the cut can be taken straight across the blank as shown, thus finishing the sides of two teeth with one travel of the cutter. When the number of teeth is even as in Figure 11.4(a), D, it is necessary to mill all the teeth on one side and then to set the cutter for finishing the oppcsite side. The maximum width of the cutter depends upon the width of the space at the narrow ends of the teeth.

When the clutch has angular teeth as in Figure 11.4(b), the cutter should be set as shown, so that on a cutter at a radial distance d equal to half the depth of the clutch teeth it lies in a radial plane.

EXAMPLE 3: MILLING SAW-TOOTH CLUTCHES

When milling clutches of this type (Figure 11.5) the axis of the clutch blank should be inclined to a certain angle α from the vertical, as shown at A. If the teeth were milled with the blank vertical, the tops of the teeth would incline towards the centre as at B, whereas, if the blank was set at such an angle that the tops of the teeth were square with the axis, the bottom would incline upwards as at C. In either case the two clutch members would not mesh completely.

For proper engagement the teeth should be cut as at D, so that the

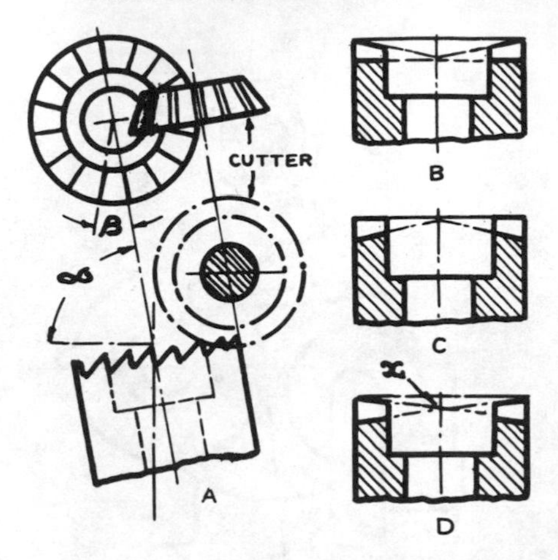

Figure 11.5
A set up for milling clutch teeth of the saw-tooth type

bottoms and tops of the teeth have the same inclination, converging at a point X. The teeth in both members will then engage across the entire width. The angle α for cutting teeth as at D can be found from

$$\cos\alpha = \frac{\sin(360^\circ/N \times \cot(\text{cutter angle})}{1 + \cos(360^\circ/N)}$$

where N is the number of teeth. If we express this formula as a rule, the cosine of the angle to which the dividing head is set equals the sine of the angle β multiplied by the cotangent of the cutter angle divided by the cosine of angle β plus one. Thus, if a saw-tooth clutch with 8 teeth is to be milled with a 60° cutter, the angle α of the dividing head setting would be

$$\cos\alpha = \frac{\sin 45^\circ \times \cot 60^\circ}{1 + \cos 45^\circ} = \frac{0.707 \times 0.577}{1 + 0.707} = 0.239 \text{ or } 76^\circ 10'$$

EXAMPLE 4: CUTTING TEETH IN END MILLS AND SIDE CUTTERS

The dividing head must be set at such an angle that the 'lands' or tops of the teeth will have a uniform width (Figure 11.6). The angle α can be found as follows.

Multiply the tangent of the angle between adjacent teeth (360° divided by the number of teeth) by the cotangent of the cutter angle. Expressed as a formula, $\cos\alpha = \tan\phi \cot\beta$, in which α is the angle of dividing head inclination, ϕ the tooth angle of the cutter blank which equals 360° divided by the number of teeth and β the angle of cutter for grooving.

Thus, if the side teeth of a straddle milling cutter are to be milled with a 70° cutter, the angle α to which the dividing head must be set will be, for 30 teeth, the angle ϕ between teeth which is $360^\circ/30 = 12^\circ$. As $\tan 12^\circ = 0.212$, and the cotangent of the cutter angle, or $\cot 70^\circ = 0.364$, therefore $\cos\alpha = 0.212 \times 0.364 = 0.077$ and $\alpha = 85\frac{1}{2}^\circ$. With an end mill the teeth on the cylindrical part are milled first, and then the end teeth are cut to match those on the periphery.

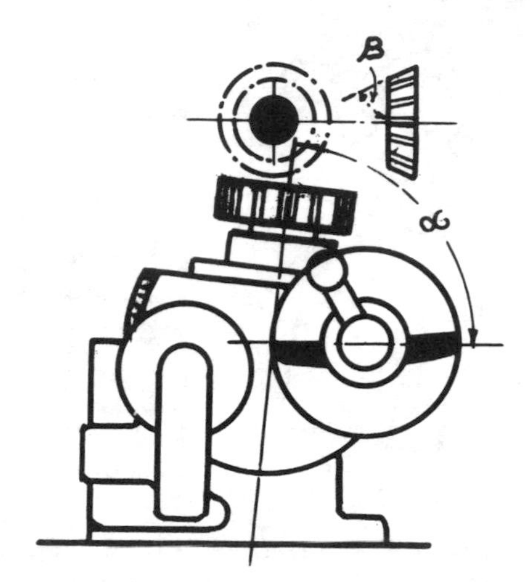

Figure 11.6.
A method of cutting teeth in end mills

EXAMPLE 5: FLUTING TAPER REAMER OR MILLING ANGULAR CUTTER

The dividing angle to which the dividing head should be set for this operation (Figure 11.7) with a 'single-angle' cutter is found as follows.

Divide the cosine of the angle θ between adjacent teeth (360° divided by the number of teeth) by the tangent of the blank angle β. The result is the tangent of angle r. Next multiply the tangent of the angle θ between the teeth by the cotangent of the grooving cutter angle ϕ, and then multiply the result by the sine of the angle r. The product is the sine of the angle δ.

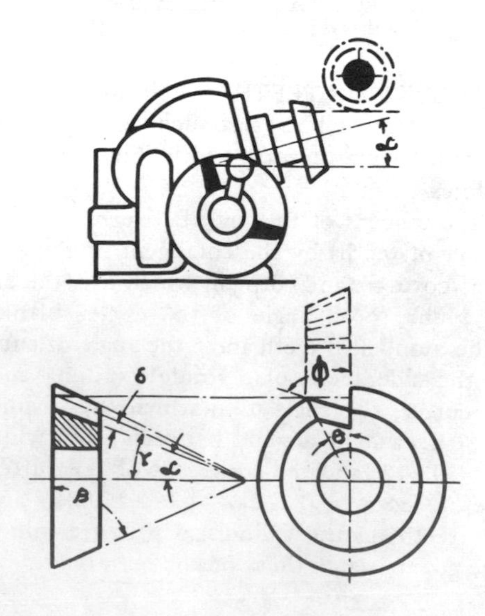

Figure 11.7.
The procedure in milling reamer flutes or an angular cutter

Subtract angle δ from angle r to obtain angle α at which the dividing head should be set. Expressing by formulae we thus have

$$\tan r = \frac{\cos\theta}{\tan\beta}, \quad \sin\delta = \tan\theta\cot\phi\sin r$$

and the angle of elevation

$$\alpha = r - \delta$$

Assume that it is desired to mill 18 teeth in a 70° (β) milling cutter blank with a 60° single-angle cutter. Then to find the angle δ to ensure 'lands' of uniform widths being milled, the angle θ between adjacent teeth is 360°/18 = 20°.

$$\tan r = \frac{\cos\theta}{\tan\beta} = \frac{0.939}{2.747} = 0.342$$

$$r = 18°53'$$

$$\sin\delta = \tan\theta\cot\phi\sin r = 0.364 \times 0.577 \times 0.327 = 0.068$$

$$\delta = 3°53'$$

Therefore the angle α at which the dividing head must be set equals the difference between r and δ or

$$\alpha = 18°53' - 3°53' = 15°$$

EXAMPLE 6: GENERATING A SPHERICAL SURFACE BY MILLING

This is an interesting experiment to generate the cap or concave surface of a true geometric sphere. One method is shown in Figure 11.8 which uses a horizontal revolving table, setting the vertical spindle of the machine to an angular position, or as in Figure 11.9 which uses the vertical head in the normal position and, by mounting the work in a dividing head, it can be tilted to the appropriate position. A single blade cutter is used, and the feed motion of either the table or dividing head is by hand rotation.

The proof of the method can be seen from Figure 11.10. If a circle is inscribed on the inside or outside of a sphere as shown in Figure 11.10(a), then the circle will touch the sphere around the circumference of the circle. If the circle and sphere are now tilted into the position shown in Figure 11.10(b), the truth of the statement is not altered. Thus, if the

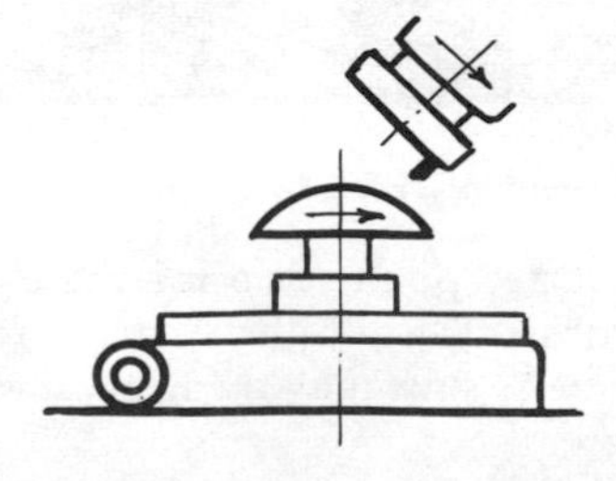

Figure 11.8. *A method of generating a spherical surface*

Figure 11.9.
The use of a dividing head for milling a cap of a sphere

circle is the locus of a cutter point, then it is merely necessary to rotate the sphere around the axis OT while the cutter is revolving, in order to generate the cap of a sphere as shown by the heavy lines in Figure 11.10(b).

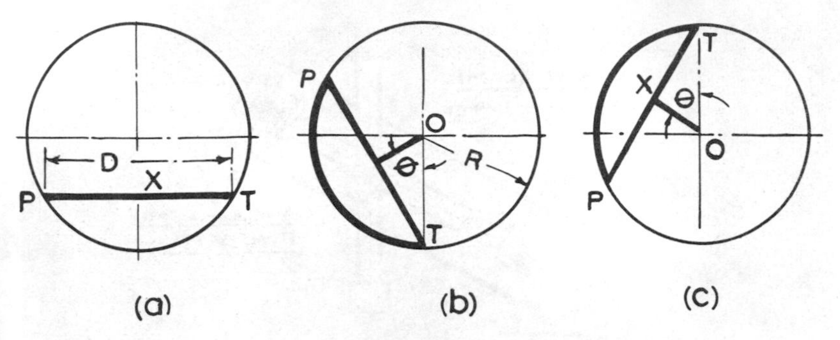

Figure 11.10.
Diagrams showing proof of the generating method

This shows the setting for a concave spherical surface, but the principle is the same for a convex surface (Figure 10.11(c)). In the latter case the cutter circle is on the outside of the sphere. In the machining operation to change over to milling a convex from a concave surface, it is only necessary to change the blank, and to set the point of the cutter on the work centre from the opposite side to that previously used.

The angle to which the dividing or vertical head must be tilted in order to generate the radius required can be found from the triangle T*O*X where *D* is the diameter of the locus of the cutter point, *R* the required radius of the sphere and θ the angle to which the cutter is to be tilted. Thus it will be seen that $\sin \theta = (D/2)R$. One practical industrial application is for machining lens grinding tools which vary from 6 to 14 in in diameter with a radius of curvature from 5 to 40 in and with a limit allowance of 0.001 in in every 10 in of curvature. Large spheres are also used for the swivel point of wheels of front-wheel drive vehicles and are often milled and ground by this method.

EXAMPLE 7: HOBBING A WORM WHEEL

This is an instructive operation in two sequences.

(1) To gash the teeth with a form cutter.

(2) To complete the wheel by hobbing.

The making of the hob can form another project and is simplified if made of high-carbon steel.

Assume that the worm wheel has 24 teeth and is required to gear with a worm of $\frac{3}{8}$ in pitch and right-hand single thread. It is of 2.864 in pitch line diameter, 3.103 in outside diameter, with a tooth depth of $\frac{3}{8} \times 0.6866 = 0.2575$ in. These dimensions would also be required for making the hob which is a replica of the worm, gashed and relieved to provide cutting edges.

First operation

Place the bronze worm wheel blank on a mandrel and mount between centres as shown in Figure 11.11. Set sectors of the dividing head for indexing the teeth; thus $40/24 = 1\frac{2}{3}$ turns. Mount cutter on the arbor,

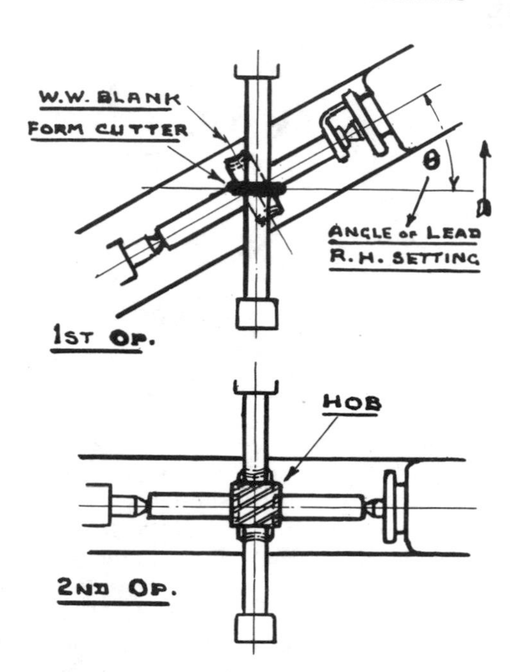

Figure 11.11. *The set-up of operations when hobbing a worm wheel*

and set central over the blank. Swivel the machine table to the angle of lead; thus

$$\text{tan of angle} = \frac{\text{lead of worm}}{\text{circumference of worm at pitch line}} = \frac{0.375}{8.993} = 0.042 = 2^\circ 25'$$

For a right-hand thread the table must be swivelled in the direction shown. Start the machine, and let the cutter touch the outside diameter of the blank. Feed to depth less 0.010 in by raising the table. Lower table, index for next tooth and continue cutting until all the teeth are gashed.

Second operation

Remove the driving dog from the mandrel so that the mandrel can revolve freely between the centres. Set table at right angles to the arbor, and replace the cutter by a hob which is set to mesh with the teeth in the blank. Start the machine, and it will be found that the hob will drive the blank and will simultaneously remove metal as it is gradually fed to full depth, so forming the teeth. (It should be noted that the plain milling machine is not provided with a swivel table, but attachments are available from machine makers and extend the range of experimental work.)

EXAMPLE 8: MILLING MACHINE FOR JIG BORING

One arrangement (Figure 11.12) is to use the tool room 'button method' of locating work where the jig buttons B are attached to the jig in positions corresponding to the holes to be bored. The jig is clamped to the angle plate A set at right angles to the spindle. There is a plug P

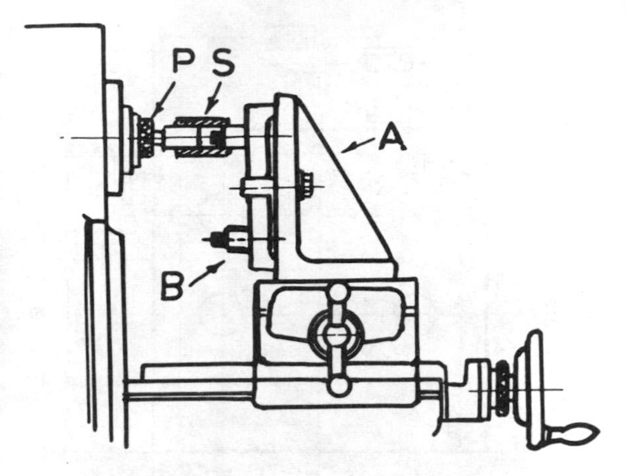

Figure 11.12.
A means of using the milling machine for jig boring:
A angle plate B jig buttons P plug S sliding sleeve

inserted in the spindle end, the end of the plug being ground to the size of the buttons. A sliding sleeve S is fitted to the plug, and, when the work is set for boring a hole, the table and knee of the machine are adjusted until the sleeve S will pass over the button that represents the location of the hole, which brings the button and spindle into alignment. A dial indicator may also be attached to the spindle nose and may be used to align the button and spindle.

All lost motion or backlash in the feed screw must be taken up. Should the button be a little higher than the spindle plug, do not lower the knee until the bush slips over the button, but lower the knee more than is required, and then raise it until the bush will pass over the button. The same applies to longitudinal movements, and, after a button has been set, it is removed, and the plug in the spindle is replaced by a drill, followed by a boring tool for finishing the hole to size. The remaining holes are machined in turn by the same method. An alternative method of setting the buttons is by using a dial indicator.

EXAMPLE 9: JIG BORING USING CO-ORDINATE DIMENSIONS

By this method the normal toleranced dimensions on the drawing are not used, but instead the operator establishes 'zero points' to each hole in order to place the work progressively under the spindle, if using a vertical head, or in front of the spindle if machining horizontally with the work fixed to an angle plate. To specify the work from hole to hole is not sufficient information, and the dimensions should emanate from one selected zero line to the left, and from another zero line from the top or bottom. The operator is then at liberty to bore the holes in any sequence that he finds advantageous.

Figure 11.13 shows an example. The dimensions enclosed by rectangles

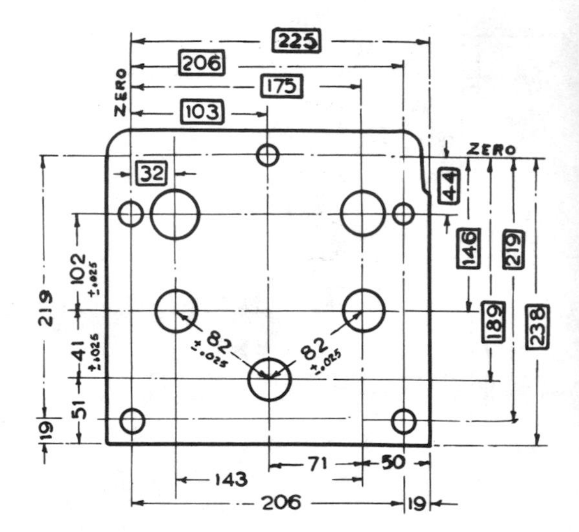

Figure 11.13.
A diagram showing the method of dimensioning a drawing for jig boring

are basic figures and do not show working tolerances. These are the dimensions the operator uses, commencing in turn from the zero point, so that the errors are not cumulative. If the problem is one of jig boring holes located equally on the circumference of a circle, the easiest way is to use a rotary table and to calculate dimensions from polar co-ordinates. It is important that the pivot of the table be located true in alignment with the machine spindle.

EXPERIMENTAL WORK ON CENTRELESS GRINDING

OBJECT

The object was to investigate the geometry of the process and to check the practical results against the theoretical calculations with respect to producing a cylindrical surface on the workpiece.

APPARATUS

The apparatus was a cylindrical grinding machine (Keighley Grinders Ltd) fitted with a Carrick centreless grinding attachment.

PROCEDURE

The attachment is shown in Figure 11.14 and comprises a bracket bolted to the table of the grinding machine. The construction comprises the regulating wheel A mounted on a spindle which is connected to a hub B by a flexible coupling C. The hub terminates in a Morse taper shank which fits into the bore of the machine workhead, and the regulating wheel is driven, via the hub, by the driving pin D of the workhead which fits into a slot. The regulating wheel is given additional support from the tailstock centre. There is a work rest W to support the workpiece, and the height of the rest can be varied by means of the adjusting knob K.

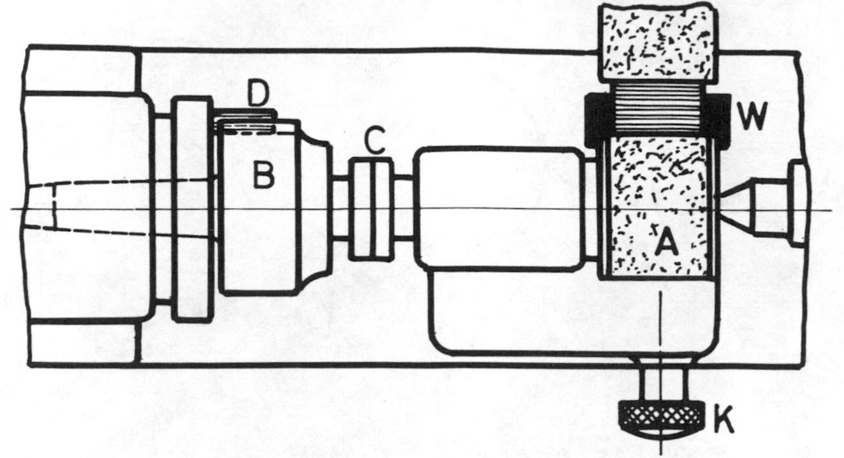

Figure 11.14.
A centreless grinding attachment for experimental work:
A regulating wheel *C flexible coupling* *K adjusting knob*
B hub *D driving pin* *W work rest*

OPERATION

Five similar workpieces were ground with the rest adjusted from a position in which all three units were on one centre line, to a height increasing by a millimetre for each workpiece. These were all slightly eccentric by the same amount, so that rounding-out could be observed. The ground specimens were measured, and the height H (Figure 11.15(a)) was indicated by considering the most accurate specimen from the trace observed from a Talyrond diagram.

Alternative tests which use cylindrical specimens with a flat or groove also showed that rounding-out would take place with the specimen above the wheel centres. A warning should be noted that the flat or groove should be small; otherwise the specimen may be trapped between the wheel and work rest and cause damage and danger.

THEORETICAL AND PRACTICAL CONSIDERATIONS

The geometry involves the following factors.

(1) The diameter of the grinding and control wheels.
(2) The diameter of the workpiece.
(3) The angle of the inclined face of the work rest.
(4) The height of centre of the work above the wheels.

The sum total of these factors can be represented by two angles, as represented by α and β in Figure 11.15(a). These two angles determine the positions of the points B and C on the workpiece, which in turn determine the magnitude and the phase, respectively, of successive irregularities on the workpiece. It follows, therefore, that the choice of the angles has a profound effect on the ultimate shape of the workpiece.

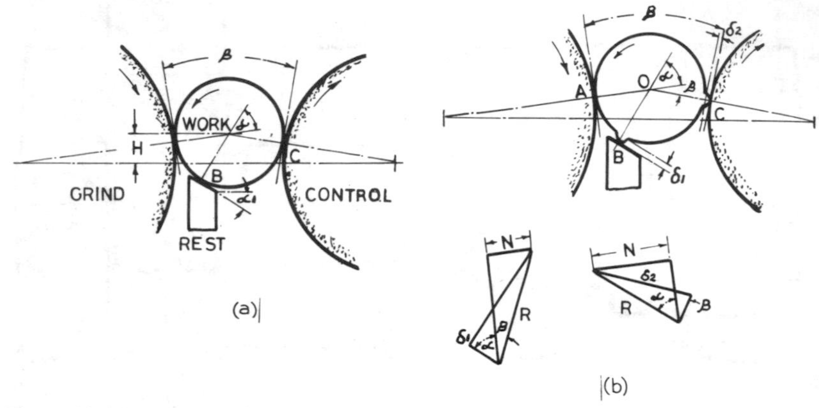

Figure 11.15.
Diagrams showing the geometry of the process

The angle α varies little for small changes in the height H, and, once a specific face angle has been chosen for the rest, the most important factor is the choice of the angle β.

From Figure 11.15(b), it will be seen that, when an irregularity on the workpiece makes contact with the control wheel, the work will be displaced by an amount δ_2, and a corresponding error will be reproduced on the work at the point of contact between the work and the grinding wheel. The position of this error will be determined by the angle β, and its magnitude by the angles α and β.

Similarly, when an irregularity on the workpiece makes contact with the rest, the component will be displaced by an amount δ_1, and the magnitude and position of the resulting error produced at the point of grind can be calculated from the angles α and β. Diagrams showing movements of the centre of the workpiece resulting from initial errors δ_1 and δ_2 are shown in the lower diagrams where N is the movement normal to wheel surface and R the resultant movement of the workpiece centre.

Investigations have shown that a work rest with a top inclined at 30° is preferable to one at 20° and that the best results are obtained when the angle β is between 6 and 8°.

EXAMPLE 10: ALIGNMENT TESTING OF MACHINE TOOLS

The instruments required for testing are relatively few and include the following.

(1) Dial gauges.

(2) Test mandrels: of these one type has a taper shank to be inserted into working spindles, and a second type to be mounted between centres.

(3) Squares: these should have as wide a base as possible to ensure accuracy and resist vibration.

(4) Spirit levels.

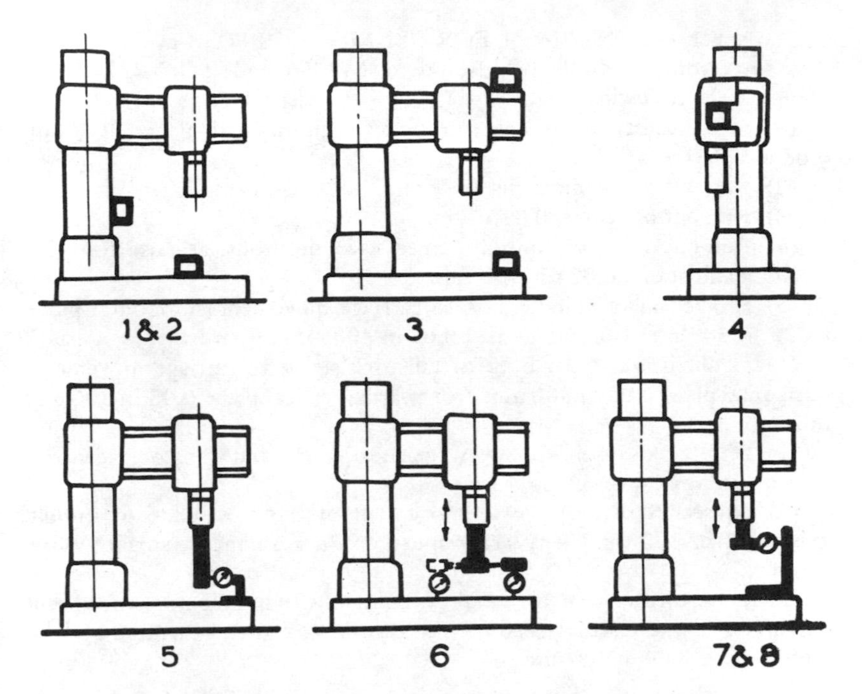

Figure 11.16.
Alignment testing diagrams for a radial drilling machine

Test charts are available for all types of machine tools, but an example of what is required for a radial drilling machine is given in Figure 11.16.

(1) Column sleeve square with baseplate in plane through centre line of baseplate (sleeve must be inclined towards the front side).

(2) Ditto in plane perpendicular to plane through centre line of baseplate.

(3) Arm parallel with baseplate (dipping at outer end).

(4) Saddle slideways, level or flat.

(5) Taper of spindle for true running.

(6) Spindle square with baseplate in plane through centre line of baseplate (spindle at lower end inclined towards column, ditto in plane perpendicular to said plane).

(7) Feed of spindle sleeve square with baseplate in plane through centre line of baseplate (inclined at lower end towards column).

(8) Ditto in plane perpendicular to said plane.

It will be noticed that if any errors exist they will be in favour of the machine during operation. Thus the arm of a radial drilling machine should dip at the lower end so that the action of drilling tends to correct this, and similar features will be noticed in the next chart for the testing of a centre lathe.

EXAMPLE 11: TEST CHART FOR GENERAL PURPOSE LATHE

The tests conform to British Standard *BS4656* and *ISO/DR1708* and comprise the following (Figure 11.17) (T.S. Harrison & Sons Ltd)

(1) Straightness of carriage movement in horizontal plane; deviation 0.03 in 2,000 mm.

(2) Run-out of spindle nose, 0.01 mm.

(3) Run-out of centre, 0.015 mm.

(4) Run-out of work spindle taper, two positions; at nose 0.01 for 300 mm and at end 0.02 for 300 mm.

(5) Parallelism of spindle axis to carriage movement; horizontal plane 0.015 in 300 mm frontwards and 0.02 in 300 mm upwards.

(6) Parallelism of taper bore of tailstock sleeve to carriage movement; horizontal plane 0.03 in 300 mm frontwards; vertical plane 0.03 in 300 mm upwards.

(7) Difference in height between headstock and tailstock centres; tailstock centre higher by 0.04 mm.

(8) Squareness of transverse movement of the cross slide to spindle axis, 0.02 in 300 mm. Any error must produce a concave surface when turning.

There are further tests for axial camming of the spindle nose and thrust bearings of the lead screw. Also for the accuracy of pitch generated by lead screw; this is 0.04 in 300 mm.

Figure 11.17.
A test chart for a Harrison lathe

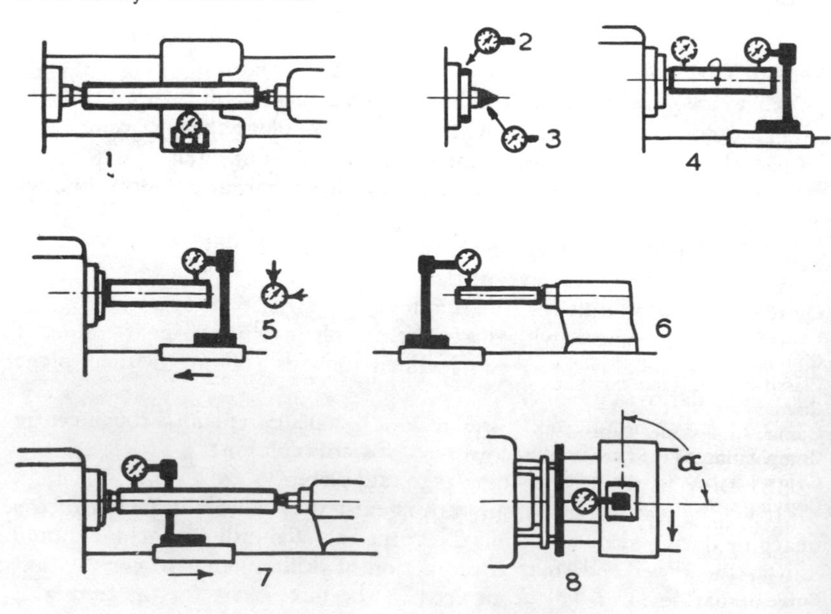

INDEX